BRITISH RAILWAYS

PAST and PRESENT

No 7

D1337948

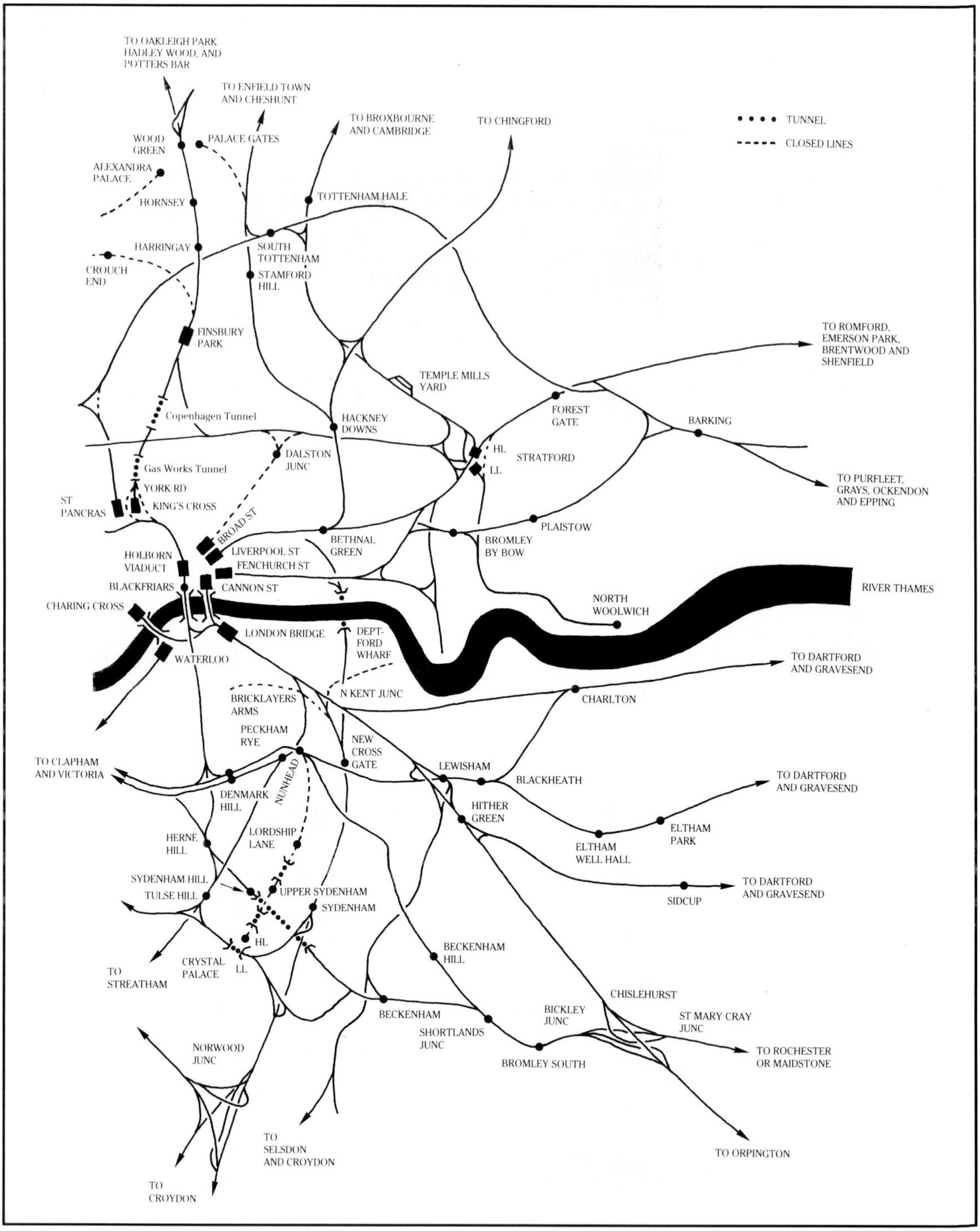
TO OAKLEIGH PARK HADLEY WOOD, AND POTTERS BAR
TO ENFIELD TOWN AND CHESHUNT
TO BROXBOURNE AND CAMBRIDGE
TO CHINGFORD
TUNNEL
CLOSED LINES
WOOD GREEN
PALACE GATES
ALEXANDRA PALACE
HORNSEY
TOTTENHAM HALE
HARRINGAY
SOUTH TOTTENHAM
CROUCH END
STAMFORD HILL
FINSBURY PARK
TO ROMFORD, EMERSON PARK, BRENTWOOD AND SHENFIELD
TEMPLE MILLS YARD
FOREST GATE
Copenhagen Tunnel
HACKNEY DOWNS
BARKING
HL
DALSTON JUNC
STRATFORD
LL
Gas Works Tunnel
YORK RD
TO PURFLEET, GRAYS, OCKENDON AND EPPING
ST PANCRAS
KING'S CROSS
BROAD ST
PLAISTOW
BETHNAL GREEN
BROMLEY BY BOW
HOLBORN VIADUCT
LIVERPOOL ST
FENCHURCH ST
BLACKFRIARS
CANNON ST
RIVER THAMES
CHARING CROSS
NORTH WOOLWICH
LONDON BRIDGE
DEPT-FORD WHARF
WATERLOO
TO DARTFORD AND GRAVESEND
N KENT JUNC
BRICKLAYERS ARMS
CHARLTON
PECKHAM RYE
NEW CROSS GATE
TO CLAPHAM AND VICTORIA
LEWISHAM
BLACKHEATH
TO DARTFORD AND GRAVESEND
DENMARK HILL
NUNHEAD
HITHER GREEN
ELTHAM PARK
HERNE HILL
LORDSHIP LANE
ELTHAM WELL HALL
SYDENHAM HILL
TO DARTFORD AND GRAVESEND
TULSE HILL
UPPER SYDENHAM
SIDCUP
SYDENHAM
HL
BECKENHAM HILL
TO STREATHAM
CRYSTAL PALACE
LL
CHISLEHURST
BICKLEY JUNC
ST MARY CRAY JUNC
BECKENHAM
SHORTLANDS JUNC
TO ROCHESTER OR MAIDSTONE
NORWOOD JUNC
BROMLEY SOUTH
TO SELSDON AND CROYDON
TO ORPINGTON
TO CROYDON

BRITISH RAILWAYS PAST & PRESENT No 7: NORTH EAST, EAST AND SOUTH EAST LONDON – This book represents a detailed examination of the changing face of the railways in the region depicted in this sketch map. The pictures have been chosen to provide a balanced view, including railways which are still in use or being developed as the book goes to press, together with scenes of railways that have been closed and either abandoned or redeveloped since the 'past' pictures were taken.

BRITISH RAILWAYS

PAST and PRESENT

No 7
North East, East and South East London

Brian Morrison & Ken Brunt

Past and Present

Past & Present Publishing Ltd

First published in February 1991
New updated edition with colour October 1996
Reprinted September 1999

British Library Cataloguing in Publication Data

A catalogue record for this book is available from the British Library

ISBN 1 85895 115 1
Past & Present Publishing Ltd
The Trundle
Ringstead Road
Great Addington
Kettering
Northants
NN14 4BW

Tel/Fax: 01536 330588
email: pete@slinkp-p.demon.co.uk

Printed and bound in Great Britain

DALSTON JUNCTION: The retaining wall of the cutting, the road bridge and the buildings overlooking the site are all that remain today of the once busy ex-North London Railway's Dalston Junction. Closed on the same day as its Broad Street terminus, 27 June 1986, the station was formerly used by Broad Street-Richmond services, which today miss out this section completely by operating direct from North Woolwich to Richmond, now a change from electric traction to diesel unit at Gospel Oak is no longer necessary. As the station footbridge is no longer in evidence, the photograph showing the present dereliction was taken from ground level in 1989; the Class '501' EMU forming the 09.57 Broad Street-Richmond train was photographed on 3 June 1979. *KB/BM*

CONTENTS

BIBLIOGRAPHY

A Regional History of the Railways of Great Britain, Vol 3 Greater London *by H.P. White (David & Charles)*
London's Local Railways *by Alan A. Jackson (David & Charles)*
London's Historic Railway Stations *by John Betjeman and John Gay (John Murray)*
BR Steam Motive Power Depots – ER *by Paul Bolger (Ian Allan)*
London Steam in the Fifties *by Brian Morrison (Ian Allan)*
Railways of the Southern Region *by Geoffrey Body (Patrick Stephens)*
Railways of the Eastern Region, Vol 1 Southern Operating Area *by Geoffrey Body (Patrick Stephens)*
LMS Engine Sheds, Vol 4 *by Chris Hawkins and George Reeve (Wild Swan)*
Great Eastern Railway Engine Sheds, Vol 1 *by Chris Hawkins and George Reeve (Wild Swan)*
30A – Steam on Stratford Shed in the 1950s *by Brian Morrison (Oxford Publishing Co)*
Jowett's Railway Atlas *by Alan Jowett (Patrick Stephens)*
Rail Atlas – Great Britain & Ireland *by S. K. Baker (Oxford Publishing Co)*
The Railmag Index *by P. J. Long (P. J. Long)*

POTTERS BAR: The A111 road bridge south of Potters Bar station needed to be completely replaced as a result of the East Coast Main Line electrification programme. Steaming under the old spans on 16 July 1955, Gresley Class 'A3' 'Pacific' No 60058 *Blair Atholl* hauls an express for King's Cross. On 22 December 1987, Class '47/0' No 47288 brings a haul of southbound 'cartics' beneath the present-day structure. *Both Brian Morrison*

INTRODUCTION

Since the railway network of Greater London is so extensive, it was agreed that two 'British Railways Past and Present' volumes would be needed in order to provide a reasonable coverage. This volume incorporates the eastern side of the capital, concentrating on the environs of King's Cross, Liverpool Street, Fenchurch Street and Broad Street, the riverside Southern Region train sheds of Charing Cross, Cannon Street and Holborn Viaduct, the 'part-termini' of Blackfriars and London Bridge, and the inner suburbs of Middlesex, Essex and Kent. The lines from Paddington, Euston, Marylebone, St Pancras, Waterloo and Victoria are covered in 'British Railways Past and Present' No 13.

The majority of the 'past' photographs are from the 1950s and are from my own collection. Of the remainder, I am indebted to my very good friend and colleague Dick Riley for allowing me the pick of his extensive collection, and thanks are also due to a few other photographers who have provided prints of subjects that were not otherwise available.

In steam days, lineside photographic permits were not difficult to obtain and these allowed vantage points from signal gantries and signal boxes that have long since disappeared. Trying to imitate a camera position of this type today can be difficult, and in a number of cases the attempt, akin to the smoking of cigarettes, can positively damage one's health! Without the steam locomotive's useful penchant for trackside fires, present-day lineside foliage is another serious problem that can be detrimental to the perfect copying of an old vantage point, although my co-author Ken Brunt's adept wielding of a machete has opened up a number of old vistas and, in some cases, has saved British Rail expenditure for lineside gardening!

The changes that have occurred in the railways of the eastern half of London over the past three or so decades have, in a few instances, been minimal. Apart from the unwelcome addition (photographically speaking) of overhead electrification catenary and supports, the King's Cross train shed, for example, remains much the same as it was in the 1950s; just outside the station, however, track rationalisation and the closure of one of the three tunnels has brought about an entirely different look from the one that was apparent only a few years ago.

On the other hand, Liverpool Street station is part of the massive Broadgate development, which has changed the area out of all recognition although the station itself has been refurbished in a most sympathetic and attractive manner. Nevertheless a number of old photographs from this venue, which would have been included herein, have had to be omitted as their present-day counterparts would show nothing more than a blank breeze-block wall!

Broad Street station has been obliterated, Holborn Viaduct is rebuilt underground and renamed first St Pauls Thameslink and later City Thameslink, while London Bridge and Blackfriars both show varying signs of refurbishment and modernisation.

Fenchurch Street, Charing Cross and Cannon Street remain basically the same internally, but with their roof exteriors drastically developed in an upward direction

Around the eastern outskirts of the capital, differences over the years are equally diverse, with complete branch lines such as that from Finsbury Park to Alexandra Palace and those to Palace Gates, Crystal Palace High Level and Gravesend West having, for the most part, returned to nature. Apart from electrification and the removal of the two-track New Barnet-Potters Bar bottleneck, the southern extremities of the East Coast Main Line remain very much the same, with parts of the old Poplar branch having now been taken over by the new and very successful Docklands Light Railway.

London steam sheds such as King's Cross, Plaistow, Devons Road, Bricklayers Arms and New Cross Gate have completely vanished from sight, while others, such as Stratford, Hornsey and Hither Green, have taken on new guises under dieselisation and electrification. Even depots built for dieselisation such as Finsbury Park have become victims of the success of modern motive power technology, where neither diesel nor electric units now need the degree of attention that once was the case. At fist left derelict, the Finsbury Park site has now been sold off and contains an attractive housing estate.

Apart from the photographers who have already been mentioned, a few words of thanks are in order to the British Rail officials who provided facilities to duplicate some of the photographs that had to be taken from within their boundary fences, and to that master printer, Derek Mercer, who has had the task of providing enlargements of nearly everything that appears in this album, both from the past and from the present. And a final word to the manufacturers of steel ladders, heavy duty denim trousers, sturdy boots, sticking plasters and insect bite ointment, without whose expertise the compilation herein would have been much less enjoyable than was the case!

Brian Morrison

DEPTFORD WHARF: A branch from Old Kent Road Junction, near New Cross Gate, opened on 2 July 1849, weaved a path beneath the London, Brighton & South Coast Railway main line and over the East London Line and made two crossings of the Grand Surrey Canal before reaching the River Thames at Deptford Wharf and Surrey Commercial Docks. Operated under the auspices of the LB&SCR and later the Southern Railway, the line lasted well into the days of British Rail and did not close finally until 1 January 1964. Today, however, it is but a memory; the Deptford Wharf swingbridge taking goods trains over the canal is now transformed into a footbridge taking pedestrians over the A206 road to the neighbouring industrial estate! On 29 March 1959, Billinton Class 'E6' 0-6-2T No 32417 crosses the Grand Surrey Canal with a freight. Almost exactly 30 years later, the scene has changed beyond all recognition. *R.C. Riley/Ken Brunt*

King's Cross

KING'S CROSS (1): Designed by Lewis Cubitt and constructed on the site of the London Smallpox Hospital, King's Cross station opened in 1852, its two great round-arched train sheds once being used, respectively, for train arrivals and departures. Apart from electrification, track rationalisation and the number of engine-spotters about, very little has changed in the 37 years that separate these particular views. (Above) On 5 July 1952, Peppercorn Class 'A1' 'Pacific' No 60122 *Curlew* awaits departure time with the 'Aberdonian' express. (Below) On 23 July 1989, Class '317/2' EMU No 317351 is stabled in the Network SouthEast suburban platforms awaiting the evening commuters while, under the station roof, InterCity 125 HSTs await their respective departure times, forming the 15.00 to Glasgow Queen Street and the 15.32 to Edinburgh. *Both BM*

KING'S CROSS (2): British Rail staff gave permission for some 40 parcels trolleys to be moved out of the way in order to copy this view of Platform 1 at King's Cross but, wisely, did not offer any assistance! (Above) On 11 January 1952, Thompson Class 'L1' 2-6-4T No 67745 prepares to haul the 18.25 semi-fast train to Baldock. The second view, taken on 27 November 1988, shows Class '317/2' EMU No 317365 forming the 15.45 service to Cambridge. *BM/KB*

KING'S CROSS (3): The once classic view of King's Cross station from atop Gasworks Tunnel was ruined when unsightly girders were erected from which to suspend the overhead electric wiring; similar electrification at next-door St Pancras station has been accomplished in a much less heavy-handed fashion. (Left) Forming the 14.30 to Leeds, an InterCity 125 HST with Class '43' power car No 43074 leading, draws away from the station on 4 February 1989. (Below) In June 1975, aesthetically happier days, Class '55' 'Deltic' No 55007 *Pinza* passes the old signal box with the 10.20 Intercity train for Leeds while two trains of empty coaching stock wait to leave, both powered by Class '31/1' diesels. *Both BM*

KING'S CROSS (4): The contrast between the old, musty King's Cross suburban platforms of Eastern Region and those of the bright, new-look Network SouthEast Sector of the present day must have a lot to do with the resurgence of commuter travel from the station. During the intervening years between these two photographs of platforms 9, 10 and 11, the latter, on the right, has been filled in and dug out again, business having first slackened and then returned to buoyancy. The above view shows a somewhat drab November day in 1972 with two Rolls Royce-engined Class '125' DMUs awaiting departure with respective services for Welwyn Garden City and Hertford North. The present-day view, below, taken on the morning of 21 January 1989, shows Class '317/1' No 317347 forming the 08.40 to Royston, and Class '317/2' No 317365 with a later departure for Hertford North. Both units are painted in the distinctive NSE livery of red, white, blue and grey. *Brian Beer/BM*

The exit from King's Cross

GASWORKS TUNNEL: (Above) Photographed from the steps of Belle Isle signal box on 27 September 1953, Class 'A3' 'Pacific' No 60067 *Ladas* climbs away from Gasworks Tunnel hauling the 15.00 King's Cross–Newcastle express. King's Cross yard is in the background and through the arches is King's Cross engine shed. With the signal box long gone and electrified wires overhead, it is not now possible to attain the same elevation for photography as can be seen from the second view, taken on 10 June 1989, of Class '91' No 91005 gliding effortlessly up the lower reaches of Holloway Bank powering the 15.10 King's Cross–Leeds Intercity service. *BM/KB*

BELLE ISLE: Looking in the opposite direction from the views on page 14, it can be seen that the viaduct carrying the North London Line still straddles the short section of permanent way between Gasworks and Copenhagen Tunnels, known as Belle Isle, and that foliage is taking over the embankment above the retaining wall, once used as rented allotments. The major change from this viewpoint, however, is the removal of the up and down tracks on which the southbound 'Plant Centenarian' special is seen (above) on 27 September 1953 passing Belle Isle signal box hauled by the preserved Great Northern Railway Class 'C2' and 'C1' 'Atlantics' No 990 *Henry Oakley* piloting No 251. The below scene shows the present-day up main line with the 12.10 Leeds–King's Cross Intercity service passing the same spot on 10 June 1989 headed by Class '43' HST/Driving Van Trailer No 43067. *Both BM*

15

COPENHAGEN TUNNEL: (Above) Bursting from the northern end of Copenhagen Tunnel on 29 June 1953, Thompson Class 'A2/1' 'Pacific' No 60508 *Duke of Rothesay* climbs towards Finsbury Park at Holloway with a morning express for Newcastle. (Below) With the two tracks in the foreground now removed, Class '91' No 91006 is seen at the same spot on 23 July 1989 with the 13.15 express from King's Cross to Leeds. *Both BM*

HOLLOWAY BANK (1): A major part of the track rationalisation that took place in the late 1970s between King's Cross and Sandy involved replacement of the original flyover at Holloway with a new one carrying realigned trackwork allowing for the utilisation of only two of the three tunnels. (Above) Heading northwards beneath the original metal structure on 11 May 1954, Class 'A1' 'Pacific' No 60114 *W. P. Allen* **hauls the 17.35 King's Cross-Newcastle express. (Below) The present-day structure is seen on 23 July 1989 as Class '91' No 91004 vanishes beneath the concrete, propelling the 10.30 Leeds-King's Cross express.** *BM/KB*

HOLLOWAY BANK (2): Viewed from the opposite side of the cutting at Holloway, the only remaining landmarks common to these two views are the trackbed remains in the foreground and the building in the background to the rear of the signal gantry; even the church has disappeared! (Above) Running on the tracks which now no longer exist, a 'Cambridge Buffet Express' prepares to dive into Copenhagen Tunnel on 11 July 1953 hauled by Class 'B17/6' 'Footballer' No 61652 *Darlington*. (Below) On 23 July 1989, today's equivalent train, the 11.50 Cambridge–King's Cross, formed of Class '317/1' EMU No 317338, is seen from the same vantage point. *Both BM*

HOLLOWAY BANK NORTH: Utilisation of multiple unit stock resulted in the extensive yards that existed on both sides of the line between Holloway and Finsbury Park becoming redundant; electric units are now serviced at Hornsey depot and the InterCity East Coast fleet is attended to at Bounds Green. (Above) Approaching Finsbury Park on the upper reaches of Holloway Bank on 22 April 1958, Thompson Class 'B1' 4-6-0 No 61302 pilots Gresley Class 'B2' 4-6-0 No 61671 *Royal Sovereign* with the 17.52 King's Cross-Cambridge/Abbots Ripton train. The present-day scene below depicts Class '317/2' EMU No 317365 forming the 14.20 Sunday service from King's Cross to Peterborough on 23 July 1989. *Both BM*

34A – Top Shed

KING'S CROSS SHED (1): The first Great Northern Railway shed was constructed on the site of 34A in 1850. Following the advent of dieselisation, it was run down and finally closed in June 1963, the locomotive allocation being transferred to New England shed (34E) and Grantham (34F) in addition to the then new diesel depot at Finsbury Park. On 9 May 1954, the line-up at the depot included Class 'A1' 'Pacific' No 60155 *Borderer,* Class 'V2' 2-6-2 No 60903 and Class 'A4' 'Pacific' No 60034 *Lord Faringdon.* After closure, the site was used for a time as a Freightliner terminal but today is mostly wasteland awaiting the overdue advent of development. *BM/KB*

KING'S CROSS SHED (2): The 70 ft high reinforced concrete coaling plant at 34A was similar to many others constructed for the LNER in the late 1920s and early 1930s and had a 500-ton bunker capacity. On 9 May 1954, Class 'A2' 'Pacific' No 60513 *Dante* fills its tender. On 10 September 1989, nothing remains to show that the concrete monolith ever existed. *BM/KB*

Great Northern suburbs

FINSBURY PARK: The track layout has changed quite considerably, the station sidings are no more and No 5 signal box has been demolished. The houses in the background remain (albeit rather obscured by advancing greenery), and the small brick-built container in the foreground, once used for junk, now makes an attempt to resemble a garden plant box. Just 17 years separate these two views of the northern approaches to Finsbury Park station, with the January 1972 one (above) showing Class '31/1' No 5545 hauling a southbound ballast train and, in the same position on 23 July 1989, Class '43' InterCity 125 power car No 43089 beginning to slow for the final lap into King's Cross, fronting the 07.25 HST from Newcastle. *Brian Beer/KB*

FINSBURY PARK DIESEL DEPOT: Following the elimination of steam locomotive working in the King's Cross area and the subsequent closure of Top Shed (34A), this country's first purpose-built diesel maintenance depot was constructed in 1960 in Clarence Yard goods depot on the west side of the Great Northern main line south of Finsbury Park station. Given responsibility for the maintenance of both main-line diesel-electric locomotives and a fleet of diesel shunters, Finsbury Park depot was closed down in October 1983 following the introduction of InterCity 125 High Speed Trains, the maintenance of which in the King's Cross area was carried out at Bounds Green, near Alexandra Palace station, the depot that now cares for the Class 91 fleet. In January 1977 (above), Class 55 'Deltic' No 55007 *Pinza* stands outside the depot with Class 08 shunter No 08551. The area remained derelict for a number of years but was later purchased for a housing development. *BM/KB*

HARRINGAY: (Above) On 15 July 1952, the 'Harrogate Sunday Pullman' titled train rounds the curve approaching Harringay station powered by Class 'A1' 'Pacific' No 60139 *Sea Eagle*. Today, the tracks are banked for high speed running, electricity abounds, in the shape of both catenary and signalling, and the down line is now the up! (Below) Propelling its Mk 3 train towards the capital on 25 May 1989, Class '91' No 91003 powers the 16.10 Leeds–King's Cross. *Both BM*

HORNSEY SHED: Primarily responsible for the Eastern Region's Great Northern line London allocation of smaller freight locomotives, Hornsey shed (34B) closed to steam traction in 1961. Without the roof ventilators, the building survives to the present time and is used as a store for the nearby Hornsey electric depot. (Above) On 3 October 1954, one of Hornsey's prolific allocation of Class 'N1' 0-6-2Ts, No 69455, stands alongside a Gresley Class 'J50/2' 0-6-0T No 68931. The view below shows how the structure looked in July 1989. *Both BM*

WOOD GREEN/ALEXANDRA PALACE: (Above) Passing through Wood Green station on 19 July 1952, a southbound fitted freight nears its destination of King's Cross Goods, uncommonly headed by Gresley Class 'A4' 'Pacific' No 60010 *Dominion of Canada*. Apart from the station name changing from Wood Green to Alexandra Palace, the present-day scene is still similar in a number of respects, although the footbridge nearest to the camera has been taken down and the other replaced with a metal structure, the old station canopies have gone and, of course, the ugly accessories to overhead electrification are everywhere. (Below) On 17 February 1989, Class '43' InterCity 125 power car No 43039 hares through the station with the 06.55 HST from Edinburgh to King's Cross. *BM/KB*

WOOD GREEN FLYOVER: The six tracks in existence in the 27 March 1954 view (above) had been reduced to four when the comparison photograph was taken on 3 June 1989, but apart from the flyover having been raised to provide the required clearance for the overhead electrified wiring, all else appears unchanged except for the motive power. In the 1950s view, Class 'A1' 'Pacific' No 60120 *Kittiwake* hauls the 13.05 King's Cross–Bradford express. The HST (below) is led by Class '43' power car No 43095 *Heaton* and is the 19.00 King's Cross–Newcastle Intercity service. *BM/KB*

WOOD GREEN TUNNEL: A gap of 35 years separates these two views of freight trains emerging from Wood Green Tunnel. The raised track on which 'WD' Class 2-8-0 No 90554 is travelling on 27 March 1954 (above) was later lowered to allow for electrification work to take place inside the tunnel, and trees and shrubs now abound on the bank and above the tunnel mouth, obscuring most of the houses which, previously, had an uninterrupted view of the East Coast Main Line. Heading southwards, hauling a variety of old wooden-sided wagons, the steam age photograph provides a contrast to the more modern image of the Railfreight Distribution Sector Class '37/0' locomotives with a haul of tanks making up the 13.10 train from Kilnhurst to Ripple Lane, Barking, on 27 July 1989 (below). The leading locomotive is No 37272 which is working in multiple with No 37059 *Port of Tilbury. Both BM*

OAKLEIGH PARK: Apart from the advent of electrification and the removal of the signal box, the environs of Oakleigh Park station would appear to have changed remarkably little in the 36 years which separate these identical viewpoints; the background poplar trees have grown to an expected extent but the two larger trees have changed not at all. (Above) On 28 February 1953, Class 'J50/3' 0-6-0T No 68968 steams through Oakleigh Park with an up freight for Hornsey. (Below) On 27 July 1989, Class '313' EMU No 313054 travels the same route, forming the 19.51 local service from Welwyn Garden City to Moorgate. *Both BM*

HADLEY WOOD: (Above) Formed of two Gresley four-car articulated 'Quadart' sets, the 18.25 King's Cross–Welwyn Garden City local service is restarted from the Hadley Wood stop on 1 July 1952 by Gresley Class 'N2/4' 0-6-2T No 69586. Shortly after this photograph was taken, Government authority was at last obtained (nothing changes!) for work to commence on opening up the tunnels on the $2^1/_2$-mile stretch of line between New Barnet and Potters Bar to enable additional up and down tracks to be laid in order to alleviate that major bottleneck on the Great Northern route. Both Hadley Wood and Potters Bar stations were modernised, with four platforms, and the works were not completed until May 1959. Following the first stage of electrification, the July 1979 view (below) shows Class '312' and '313' EMUs forming, respectively, a King's Cross–Royston semi-fast train and a Moorgate–Welwyn Garden City service. *Both BM*

HADLEY WOOD NORTH TUNNEL: Track maintenance personnel were plain 'gangers' when the above photograph was taken, on Sunday 13 July 1952, of Thompson Class 'B1' 4-6-0 No 61027 *Madoqua* ambling away from Hadley Wood North Tunnel with an up freight for King's Cross Goods. (Below) On 4 November 1989, Class '317/1' EMU No 317345 is travelling very much faster, forming the 14.10 train from Cambridge to King's Cross. Four tracks, electrification and colour light signalling are now apparent at this previously rural setting. *Both BM*

POTTERS BAR (1): (Above) Prior to the East Coast Main Line widening that took place during the 1950s, Gresley Class 'A4' 'Pacific' No 60007 *Sir Nigel Gresley* climbs the gradient approaching Potters Bar on Sunday 28 September 1952, hauling the 09.15 King's Cross–York 'Centenarian Express' special which was run to commemorate, among other things, the opening of the London terminus. (Below) At the same spot, on Sunday 23 July 1989, the 08.50 local service from King's Cross to Welwyn Garden City is formed of Class '317/2' No 317365 in the livery of the Network SouthEast Sector. *Both BM*

POTTERS BAR (2): Three views of Potters Bar cutting taken on 12 May 1951, 16 July 1955 and 17 November 1989: (Top) Prior to the widening, Class 'N2/2' 0-6-2T No 69490, steaming freely, hauls two sets of 'Quadart' articulated coaches on a stopping service for Hatfield, indicated by the route destination board fitted. (Centre) Gresley Class 'V2' 2-6-2 No 60849 rasps up the gradient with a northbound Anglo-Scottish fitted freight soon after the cutting was widened to four tracks, but prior to the new tunnel bores being completed to enable the four track section to extend further south. (Bottom) The latest technology for the ECML, with the southbound 'Yorkshire Pullman' consisting of all Mk 4 stock headed by Driving Van Trailer No 82203, with Class '91' No 91009 propelling at the rear. *All BM*

The North London line

PALACE GATES: (Above) On 10 April 1954, a train for North Woolwich waits to leave Palace Gates station hauled by Holden Class 'F5' 2-4-2T No 67209. This station, near Wood Green, was closed in January 1963 when the spur from South Tottenham was abandoned and North London line services were extended to Gospel Oak. Contained within Bounds Green traction depot yard, a very small section of the platform end, furthest from the camera in this view, still remains today, but the whole of the station buildings have been totally eradicated by the construction of a modern housing estate, the present-day view (below) having been taken from the same standpoint in February 1989. *BM/KB*

STRATFORD (LOW LEVEL): The 12.45 Palace Gates-North Woolwich train pauses at Stratford (Low Level) station on 8 August 1953, headed by Class 'F5' 2-4-2T No 67219 running bunker first. Today (below), the eastbound platform has been extended, and a brick wall erected on the left has narrowed the westbound one. Behind Class '313' EMU No 313021, high fencing has been erected, and the dominant three-storey brick building has been demolished. *BM/KB*

FORK JUNCTION, STRATFORD: On 16 February 1957 Thompson Class 'L1' 2-6-4T No 67701 passes Fork Junction box as it restarts a North Woolwich-Palace Gates train from Stratford (Low Level) station and runs beneath the Liverpool Street-Norwich main line above. (Below) On 5 August 1989 the 14.35 North Woolwich-Richmond service is formed of Class '416/3' 2EPB unit No 6318; there is no Fork Junction signal box to pass, however, or even a Fork Junction, the line that forked off to Temple Mills having been closed in June 1969 and subsequently taken up. The motive power was later returned to the Southern Region, being replaced by Class '313s'. *R. C. Riley/BM*

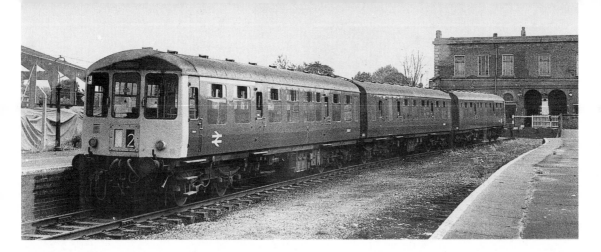

NORTH WOOLWICH: The arrival platform at North Woolwich station is now given over to the Great Eastern Museum where displays from the Passmore Edwards Museum Trust are housed both inside and outside the old station building. (Top) On 10 August 1973, however, that was the only platform in use, with the tracks on the departure side having been taken up. The three-car Birmingham R&CW Class '104' DMU forming the 12.54 train to Stratford (Low Level) was withdrawn in 1981-82. (Above) Sixteen years later to the day, the departure platform has been reinstated and is now the only one in use for British Rail services, a view of the Museum yard being possible over the fence which has been erected. The Class '416/3' EMU forming a service to Richmond is No 6308. *BM/KB*

(Left) The original 1847 Italianate North Woolwich station building was almost derelict when photographed on 6 April 1983 but, thanks to a sponsorship from the London Dockland Development Corporation, the building has since been completely restored. The view of what is now the Great Eastern Museum frontage was taken on 10 August 1989. *BM/KB*

The Poplar Docks line

DEVONS ROAD, BOW: Once boasting a North London Railway train every 15 minutes from Broad Street, bomb damage sustained by London's Docklands during the Second World War, coupled with a general decline in revenue, resulted in cessation of passenger services to Poplar from 14 May 1944. Tracks east of Millwall were taken up in 1967 and a direct line constructed from the NLR yard at Poplar to the western Poplar Docks across the bed of the erstwhile Blackwall Railway. This line replaced the old high-level ex-NLR access and remained in occasional use until the advent of the Docklands Light Railway. (Above) In connection with the then-projected Docklands Light Railway, a Cravens two-car DMU, consisting of motor brake and driving trailer cars Nos E53362/54420, was used on the line on 9 February 1984 to ascertain whether a regular and frequent train service would be likely to have any adverse noise or vibration effect upon the local buildings or their inhabitants in the event of an intensive train service coming about. The test train operated on the line for some five hours and is seen above near Devons Road, Bow. The present-day scene from the same viewpoint (right) shows a DLR driverless train approaching the Docklands station of Devons Road with the tell-tale high-rise flats still showing up in the background, although now partly obscured by the new building under construction on the left. *BM/KB*

DEVONS ROAD SHED: With some of the flats shown opposite under construction, the view (above) of Devons Road engine shed yard on 16 April 1955 shows Ivatt Class '4MT' 'Mogul' No 43001 as originally constructed with a double chimney. Opened by the North London Railway in 1882, Devons Road (1D) closed for steam traction in 1958 and became the first all-diesel shed in Great Britain; final shutdown came about in 1964. When the more recent view was taken in 1989, the area was derelict and only the background flats remained. However, it has since been developed and is now an industrial complex. *BM/KB*

The 'Ally Pally'

ALEXANDRA PALACE: The Great Northern Railway branch from Highgate to Alexandra Palace, known as the 'Ally Pally' line, opened with the Palace itself on 24 May 1873 and initially brought great crowds. However, the fluctuating fortunes of the Palace following its destruction by fire on 9 June 1873 brought about no fewer than eight closures and re-openings of the line between 1873 and 1898, and it eventually closed for the last time on 5 July 1954. (Above) On 10 April 1954, the 14.45 auto-train for Finsbury Park leaves the terminus propelled by Hill Class 'N7/2' 0-6-2T No 69694. Behind the trees to the right of the present-day view can just be seen the Palace mast. *BM/KB*

CROUCH END: (Above) With the Class 'N7/2' No 69694 now leading the push-pull set and seemingly exuding steam from every crack in the bodysides, the 16.02 Finsbury Park–Alexandra Palace train restarts from the Crouch End stop on the same day. As can be observed from the present view of this location, photographed in February 1989, the old trackbed is now a footpath. *BM/KB*

The Southern's Thames-side termini

CHARING CROSS: For many years nothing very much occurred to alter the appearance of Charing Cross station from its Southern Railway days, even the 'SR' initials and coat of arms being retained under Nationalisation. In the late 1980s, however, the value of the air space above the station platforms was realised by the BR Property Board, the result being the massive development shown below on 5 September 1995 - happily retaining a revised form of the original 'SR' insignia. The train departing is formed of Class '465/2' No 465211, one of the new 'South Eastern' 'Networkers' which have completely replaced the old 4EPB slam-door EMUs that filled all six platforms in the photograph dated 3 January 1959. Following completion of Hungerford Bridge, the original Charing Cross station was opened by the South Eastern Railway in 1864 and, because of its location near Trafalgar Square, was considered as the most illustrious in London. Its fortunes have since risen and fallen with those of the Strand, at the western end of which the station stands. *R. C. Riley/BM*

CANNON STREET (1): (Above) On 25 April 1951, some eight years prior to the removal of the overall station roof, the fireman of Maunsell Class 'N15' 'King Arthur' 4-6-0 No 30806 *Sir Galleron* (the last of the 'Arthurs' to be built) has checked the coupling ready for an evening departure to Hastings. (Below) Seen from the same position on 7 June 1976, empty EPB stock awaits the 'crush hour' of 17.00 and the daily onslaught of City commuters. *Both BM*

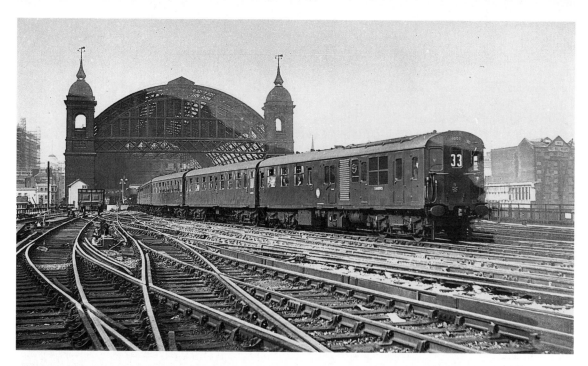

CANNON STREET (2): Apart from the 120-foot twin baroque towers, little remains today of the original 1866 South Eastern Railway terminus, Sir John Hawkshaw's great crescent roof having been dismantled in 1958/9 as part of extensive station rebuilding. The present climate of preservation would probably not have allowed removal to have taken place, despite the official reasons given at the time that the roof was unsafe due to a combination of age, corrosion and wartime bomb damage all taking their toll. Cannon Street station was constructed to serve the City of London, and that is still its use over 130 years later. An estimated 75,000 City commuters use the station during the weekday morning and evening peaks, when trains run to and from all the expected suburban conurbations of South East London and Kent. Off peak, a 15-minute shuttle service operates from London Bridge to supplement a much-reduced timetable, and at weekends the terminus is normally closed. The scene from the past shows the inaugural service to Hastings by diesel-electric multiple units Nos 1003/4 on 6 May 1957. Currently, a modern office complex straddles the station between the towers as Class '465/1' 'Networker' No 466163 departs forming the off-peak shuttle to London Bridge. *Both BM*

HOLBORN VIADUCT (1): Holborn Viaduct station opened in March 1874, followed by Snow Hill (later Holborn Viaduct Low Level) in August of the same year. Both provided much-needed relief to commuters who previously had been required to use the congested island platforms at Ludgate Hill station (closed in 1929), situated on the through line to Farringdon. Although closed in 1916, the Low Level platforms remained visible just inside Snow Hill tunnel, which was kept open until 1969 for through freight traffic, thereafter being out of use until the link to Farringdon was restablished for the introduction of 'Thameslink' services in April 1988. In order to gain access to the tunnel, the 'Thameslink' trains had to bypass the high-level Holborn Viaduct platforms, and that station ultimately closed in 1990. It was replaced by St Pauls Thameslink (later renamed City Thameslink), located near the site of the old Low Level platforms. On 18 August 1989 (above), the 15.38 Sevenoaks-Bedford 'Thameslink' train, formed of Class '319' EMU No 319019, enters Snow Hill tunnel, while trains to Dartford via Sidcup and to Canterbury East await departure time in Holborn Viaduct station alongside.

(Below) A massive engineering project has obliterated the high-level platforms, allowing the tracks into Snow Hill tunnel to be slewed to the right to gain access to Blackfriars station via a remarkable 1 in 29 gradient. On 29 May 1990, the same Class '319' unit forms the 08.50 Luton-Sevenoaks train, approaching the City Thameslink stop. Then the offices of Williams National House could still be seen above the remains of the Holborn Viaduct concourse, but today this scene is now underground, beneath new office complexes that have been constructed over the railway. *Both BM*

HOLBORN VIADUCT (2): For a pleasant change, it is possible here to show the 'past' photograph as one of overgrown trackbed and the 'present' one with the railway restored. (Above) On 13 September 1982 the empty stock of Class '415/1' 4EPB No 5196 awaits the evening commuter business while the ground to the left appears as if it may be suitable for a picnic. (Below) On 18 August 1989, however, not only is the line and tunnel restored, but entirely new business complexes have replaced the old order. Class '319' No 319044 heads into the tunnel and its next stop at Farringdon, forming the 15.20 Sevenoaks-Luton 'Thameslink' service. Following the closure of Holborn Viaduct station in 1990, the complete raft from where these photographs were taken was removed together with the area containing the platforms, and the track at this point was slewed to the right to travel beneath the old station to reach the new City Thameslink station below. It then rises on a remarkable 1 in 29 gradient to reach **Blackfriars.** *Both BM*

BLACKFRIARS: Originally named Blackfriars Bridge, the first station opened in 1864 on the east side of the approach to the new road bridge over the Thames then being constructed. The terminus then moved over the river to another short-lived site in Little Earl Street. Thereafter the opening in 1865 of Ludgate Hill station catered for passengers until St Pauls station opened on the present-day site in 1886; its name was changed to Blackfriars when St Pauls Underground station was opened in 1937. (Above) On a very murky 10 October 1954 Class 'T9' 'Greyhound' 4-4-0 No 30729 is the subject of attention from enthusiasts as it waits to leave one of the three terminus platforms with what is described on the headboard as an 'Inter-Regional Ramblers Special'.

(Below) Completely rebuilt between 1972 and 1977, the present-day Blackfriars is probably busier than at any time in its long history, with through all-day 'Thameslink' services from Bedford, Luton and St Albans to Brighton and Sutton, and trains from the terminus platforms to Orpington and to Dartford via Bexleyheath. On a sunny 9 November 1991 Class '319' No 319029 restarts the 12.00 Luton-Guildford 'Thameslink' train from the southbound through platform, while in the three terminus platforms are Class '33/1' No 33114, with a charter to Sevenoaks, and two stabled 4EPB units. *R. C. Riley/BM*

LONDON BRIDGE (1): Two views of trains departing from the Central Section terminus of London Bridge graphically illustrate something of what has occurred to London's city skyline in the intervening 30 years; just one building, on the right, remains to show that the locations are the same. (Above) On 14 May 1959, Wainwright Class 'D1' 4-4-0 No 31735 draws away from the station with the 12.44 vans train for Ramsgate. (Below) On 18 August 1989, the 10.19 Sanderstead–London Bridge train enters the station formed of Class '423/1' 4VEP EMU No 3465; 4EPB and 4CIG stock is stabled awaiting the evening commuters and another 4EPB enters the Eastern Section of the station with a local service to Charing Cross. *R. C. Riley/BM*

LONDON BRIDGE (2): The half-barrel-shaped cast iron roof of the London, Brighton & South Coast Railway terminus of London Bridge has survived despite the ravages of time and, in particular, Adolf Hitler's bombs. (Above) On 14 May 1959, Billinton Class 'E4' 0-6-2T No 32474 steams away from beneath the roof hauling a short vans train bound for New Cross Gate. (Below) As can be observed from the similar view taken on 13 September 1982, the platforms have been lengthened and the whole station is now in the shadow of the multi-storey Guy's Hospital building to the rear. *R. C. Riley/BM*

LONDON BRIDGE (3): With trackwork on the 'South Eastern' side of the station now raised above the level of the Network South Central platforms, and the platform itself relocated and lengthened, only the wall of the station terminus now shows that these two views are taken from the same position. (Above) On a dull 23 June 1951, Marsh Class 'I3' 4-4-2T No 32091 is watered as a 4EPB unit passes in the background on Eastern Section lines. *Both BM*

LONDON BRIDGE (4): The station refurbishment completed at London Bridge in the 1980s was no more than the old buildings deserved, having only been patched up in one manner or another during the years following extensive damage sustained during the London Blitz. As London's first railway terminus, it was opened in 1836 as a simple two-platform structure for the London & Greenwich Railway, which ran to Greenwich across brick arches. The view of Ivatt Class '2MT' 2-6-2T No 41300 running through Platform 2 on 13 March 1957 (above) contrasts with the new-look premises of 18 August 1989 (below), where can be seen the new platform canopies and passenger overbridge, if not the fresh paint. At Platform 2 is Class '415/1' 4EPB No 5126 awaiting custom as the 11.19 shuttle to Cannon Street. *R. C. Riley/BM*

South London scenes

NORTH KENT WEST JUNCTION: (Above) Passing the signal box at North Kent West Junction on 29 March 1958, Maunsell Class 'W' 2-6-4T No 31919 heads a transfer freight to Hither Green. Until Bricklayers Arms shed was closed in 1962, all movements in and out had to made by way of this busy junction which, on 24 April 1989, still carried track slowly being hidden by the spreading undergrowth (below). Also still remaining in situ on the site was the base of the old signal gantry on the left. *R. C. Riley/KB*

BRICKLAYERS ARMS SHED: The terminal station of Bricklayers Arms was jointly opened by the South Eastern Railway and the London & Croydon Railway in 1844 as a means of avoiding the tolls imposed by the London & Greenwich Railway for the use of its metals at London Bridge. The description of the station as a new 'Grand West End Terminus' would, today, have brought proceedings under the Trades Descriptions Act! The hoped-for passenger traffic did not materialise in sufficient volume for the venture to continue, and the two-road structure closed in 1852 and became the original Bricklayers Arms engine shed. The depot became the heart of the Bricklayers Arms goods complex and maintained locomotives used for passenger traffic from Charing Cross and Cannon Street to the Kent coast as well as myriad goods engines. The above view, taken on 2 April 1955, shows Class 'L1' 4-4-0 No 31788 alongside Class 'E4' 0-6-2T No 32564 outside the four-road extension to the complex which dated from 1865, albeit re-roofed by the Southern Railway since that time. 73B was closed in 1962 and a parcels depot was opened there in 1969, the site eventually being abandoned in 1981. (Below) Purchased from BR for much-needed housing development in the area, this was the view of the site on 29 July 1989 taken from the same standpoint. *Both BM*

NEW CROSS GATE: One platform of the ex-London, Brighton & South Coast Railway station at New Cross Gate (originally New Cross) is given over to the East London Line of London Underground where services run to and from Whitechapel. (Above) On 3 June 1956, the platform was used for a special ramblers' train, the 'John Milton Special', which travelled from there to Chesham with a London Transport locomotive on each end, there being no run-round facilities at the Buckinghamshire terminus for the return journey. With No 14 *Benjamin Disraeli* waiting to leave, there would seem to be little that has changed here over the years apart from a reduction in the length of the platform canopy and the hut having been painted white! (Below) On 28 December 1988, refurbished LT A60 stock, with Driving Motor Car No 5066 leading, waits to depart with the Whitechapel shuttle. *Both BM*

NEW CROSS GATE SHED (1): The complicated history of the rambling buildings that made up New Cross (later New Cross Gate) engine shed began in June 1839 when the London & Croydon Railway opened a roundhouse there. Under the auspices of the LB&SCR and later the Southern Railway, the depot was extended on a number of occasions and for many years was important as the Brighton line's main London depot. Electrification and the transfer of maintenance to other depots resulted in closure in 1957, although no locomotives were allocated there after 1949 due to the buildings being open to the elements as a result of bomb damage sustained during the war. 'New shed' was constructed in 1929, although it looked anything but new on 23 June 1951 (above) with Billinton Class 'E3' 0-6-2T No 32459 simmering outside. Apart from one outside wall, nothing is left standing of the old complex today, the view below being taken from the same position on 28 December 1989. *BM/KB*

NEW CROSS GATE SHED (2): (Above) Where Bulleid's unsteamed second 'Leader' Class 0-6-6-0T No 39002 was dumped on 23 June 1951, an oil storage tank stands today. On the day following the photograph being taken, the locomotive was towed to Brighton Works and cut up. *Both BM*

HITHER GREEN SHED: Primarily concerned with freight locomotives, Hither Green shed was opened by the Southern Railway in 1933 and closed to steam in 1961. (Above) On 2 May 1959, new Class '24' diesel-electric locomotives are lined up at 73C alongside the steam engines that they were intended to replace, and which include a variety of Maunsell 'Moguls', Wainwright 'C' Class 0-6-0s and a Bulleid Class 'Q1' 0-6-0. (Below) Converted for diesel traction, in 1985 the shed was downgraded to a locomotive stabling and fuelling point, but is currently seeing a new lease of life as a Mainline Freight Ltd depot and a base for track maintenance machines. From atop the same lighting mast, the photograph taken on ASLEF strike day, 14 July 1995, shows the completion of the two-road fuelling shed (which in 1959 had received only its concrete base) and demolition of the right-hand side of the main shed building with a new office building constructed to the rear. The remaining part of the old shed is now occupied by 'Plant South East', and the tower of Canary Wharf can be glimpsed in the distance. Classes '58' and '60' now dominate. *R. C. Riley/BM*

DENMARK HILL (1): On 14 May 1959, Maunsell three-cylinder Class 'N1' 'Mogul' No 31876 scurries through Denmark Hill station with an excursion from Victoria to Ramsgate and contrasts with the scene below, taken on 9 August 1982, of Class '415/1' 4EPB No 5220 in the station and forming the 16.50 Holborn Viaduct–Bellingham local stopping train. King's College Hospital and its laundry chimney now dominate the background, only some rubble marks the position of the signal box, and the station building has suffered fire damage, the top portion having been destroyed. This was later to be rebuilt, however, and today bears more resemblance to the steam age view than the one from 1982. *R. C. Riley/BM*

DENMARK HILL (2): Stabling facilities for locomotive-hauled carriage stock in the 1950s were not over-plentiful, and often rather out-of-the-way sidings had to be utilised during the daytime to await morning or evening peak services. (Above) The 15.40 empty coaching stock working from Blackheath yards in South East London, to form the 18.14 Cannon Street–Ramsgate, is taken through Denmark Hill station on 6 May 1959 by Wainwright 'C' Class 0-6-0 No 31717; the uncommon disc code signifies a North Kent Lines train via Nunhead. When the comparison photograph was taken on 25 November 1989, embankment foliage precluded exactly the same footsteps being used to record the 10.51 Sevenoaks–Cricklewood service, formed of Class '319' EMU No 319007, but little appears to have changed apart from demolition of the building on the embankment which was once a sub-station for the LB&SCR overhead electrified lines. *R. C. Riley/BM*

PECKHAM RYE (1): (Above) Passing Peckham Rye coal sidings, north of the station, on 10 October 1953, two old-type 4SUB EMUs, led by No 4207, form a London Bridge to London Bridge local circular service travelling via Forest Hill, Crystal Palace and Tulse Hill. (Below) On 29 July 1989, the 11.12 train from West Sutton to London Bridge is formed of Class '455/8' EMU No 5802 and passes the same point; the coal sidings have long since disappeared and the track on which the train is travelling has been slewed. The old warehouse on the left of the scene, which rather looked as if it was likely to fall down at any minute, in fact still survives to the present day. Looks are not everything! *R. C. Riley/KB*

PECKHAM RYE (2): (Above) On 28 February 1957, Class 'W' 2-6-4T No 31923 hauls a freight for Hither Green past the Peckham Rye depot built in 1908-9 by the LB&SCR for maintenance and repair of its South London line overhead electric stock. (Below) On 29 July 1989, the 10.42 Cricklewood–Sevenoaks 'Thameslink' service, formed of Class '319' No 319048, passes the same spot. An unusual-looking gathering of new dwellings now occupies the depot site, and on the opposite side of the line the old houses have gone, converted into the Warwick Gardens parkland, a site proposed for the emergence into daylight of the underground high-speed rail link for the Channel Tunnel! *R. C. Riley/BM*

HERNE HILL: Apart from the introduction of multiple aspect signalling, very little appears to have changed at Herne Hill over the years. (Above) Hauled by Riddles BR Standard '7MT' Class 'Britannia' 'Pacific' No 70014 *Iron Duke*, the southbound 'Golden Arrow' express from London Victoria to Dover Marine passes the station on 20 October 1957 formed of all-Pullman stock. (Below) On 19 August 1995, Class '373' 'Eurostar' Nos 3227/8 pass forming the 16.23 train from Waterloo International to Paris Nord. *R. C. Riley/KB*

Past and Present Colour

North East, East and South East London

HORNSEY: Hauling a typical King's Cross local service of the day, Class 'N2/2' 0-6-2T No 69526 passes Hornsey signal box on 20 September 1958.

On 31 August 1995, only the long station footbridge remains, as a current King's Cross local service, formed of Class '313' EMU No 313032, passes the same spot heading for Hertford North. The old station building has disappeared together with Hornsey box, and the trappings of electrification are everywhere. *R. C. Riley/Brian Morrison*

WOOD GREEN/ALEXANDRA PALACE: Approaching Wood Green on 13 September 1958, a long fitted freight is hauled northwards by BR Standard '9F' Class 2-10-0 No 92193.

Currently the bridge parapet and gas holder survive, as the 12.46 European Passenger Services HST from Waterloo to Edinburgh approaches what is now Alexandra Palace station, headed by Class '43' DVT power car No 43080, with No 43087 bringing up the rear. *R. C. Riley/Ken Brunt*

BETHNAL GREEN: Passing the platforms on 28 February 1959, the last-built 'K1' Class 'Mogul', No 62070, hauls empty coaching stock from Liverpool Street to Thornton Field carriage sidings, a location that has become pluralled to Thornton Fields over the years.

The station buildings are now demolished, and an additional catenary support has been erected, as the 11.30 Liverpool Street-Norwich InterCity Anglia express passes the same place on 31 August 1995, led by DBSO No 9707 and powered from the rear by Class '86/2' No 86232 *Norwich Festival***, since renamed** *Norfolk and Norwich Festival. R. C. Riley/Ken Brunt*

NOEL PARK: The small intermediate station between West Green and Palace Gates on the North Woolwich to Palace Gates line was the location chosen for an exhibition between 12 and 14 September 1958 in connection with the Borough of Wood Green's charter celebrations. Some 14,000 people attended, and among the exhibits on display can be seen 'J52/2' Class 0-6-0ST No 68846 (now preserved as GNR No 1247) and 'C12' Class 4-4-2T No 67352, both in ex-works condition. Also in the frame is the first of what became a Class '26' diesel, No D5300.

The same view in August 1995 reveals only the Parish Church of St Mark to identify the location. *R. C. Riley/Ken Brunt*

CHISLEHURST: Rebuilt 'West Country' 'light Pacific' No 34005 *Barnstaple* passes Chislehurst Goods signal box on 16 May 1959, hauling the 12.53 Charing Cross-Folkestone express.

On 31 August 1995 the 17.13 Charing Cross-Margate train travels along the same stretch of line, formed of Class '411/5' 4CEP No 1534 leading Class '423/1' No 3570. The goods sidings have been lifted, the signal box and attendant semaphore signals are gone and lineside trees completely obscure the houses below the embankment. The colour light gantry still remains but has been moved to a position nearer to the camera. At least the present-day motive power appears much cleaner. *R. C. Riley/Brian Morrison*

ST MARY CRAY JUNCTION (1): Passing the junction, between St Mary Cray and Bickley stations, on 16 May 1959 is 'King Arthur' Class 4-6-0 No 30769 *Sir Balan* heading for the Kent Coast with a Victoria-Ramsgate train.

The same footbridge from which the steam days view was taken is still there for the present-day photographer, and the trackwork at the junction is unchanged. What makes such a contrast is the considerable growth of trees and bushes on both sides of the line, almost changing the appearance of the location out of all recognition. On 29 July 1995 Class '411/5' 4CEP No 1511 passes, forming a Victoria to Canterbury West train. *R. C. Riley/Brian Morrison*

ST MARY CRAY JUNCTION (2): Looking in the opposite direction, we see 'Schools' Class 4-4-0 No 30915 *Brighton* approaching the junction on the same day, hauling a London-bound express from Ramsgate.

At the same location today, the up and down lines have been changed over, and it is now a train travelling in the opposite direction that is required in order to obtain a similar view. Class '465/2' 'Networker' No 465236 travels the same track on 29 July 1995, forming the 14.13 Blackfriars-Sevenoaks train. *R. C. Riley/Ken Brunt*

TULSE HILL TUNNEL: One of the last workings of a 'Remembrance' Class 4-6-0 was the 'Riverside Special' railtour from London Bridge to Windsor on 23 June 1957 behind No 32331 *Beattie*, **seen bursting out of Tulse Hill Tunnel.**

Today the tunnel portals appear in better condition than nearly 40 years ago, but the same cannot be said for the view, now almost hidden behind a mass of brambles covering the lineside embankment. *R. C. Riley/Brian Morrison*

TULSE HILL: (Above) Prior to having a headcode panel fitted and all front-end lighting removed, Raworth & Bulleid Class '70' Co-Co electric locomotive No 20003 passes Tulse Hill station on 7 November 1964 hauling a London-bound excursion. In the view above, the gas lamps have been replaced by electric ones, the station nameboard is vintage Network SouthEast and, of course, the semaphore signals have all gone. The 13.13 Purley to Luton 'Thameslink' service slows for the station stop on 24 April 1989 formed of Class '319' unit No 319036. *R. C. Riley/KB*

CRYSTAL PALACE (1): (Above) On 23 June 1954, a Willesden-allocated Fowler Class '3MT' 2-6-2T No 40006 emerges from Crystal Palace tunnel with the 09.35 Willesden to East Croydon van train. (Below) With no sign remaining of the Crystal Palace Tunnel signal box and with foliage now cascading down the brickwork, the 13.06 service from London Victoria to West Croydon passes the same spot on 29 July 1989 formed of Class '455/8' EMU No 5803. *R. C. Riley/BM*

CRYSTAL PALACE (2): (Above) Bound for Norwood Junction yards, Class 'W' 2-6-4T No 31914 takes a train of mixed wooden-sided and steel wagons through the 'West station' at Crystal Palace on 12 March 1954. (Below) With the tall BBC TV aerial now in the background and the old roof structure the victim of official vandalism, Class '455/8' No 5815 makes the scheduled stop beneath the new platform canopies on 29 January 1989 forming the 12.21 Victoria–West Croydon service. This part of the Crystal Palace station complex is by far the busiest, served by local trains from both Victoria and London Bridge in addition to 'Thameslink' services and a variety of freight traffic. *R. C. Riley/BM*

CRYSTAL PALACE (3): The platforms at Crystal Palace 'East station' are little used today apart from a few peak-time trains to and from London Bridge via Sydenham. As a precaution, the overall roof was removed following collapse of the similar one at Charing Cross in 1905; the two bay platforms and two centre sidings (between which was once an engine pit) have been taken up, together with the tracks on the left leading to sidings. The view taken on 23 June 1954 (above) shows three types of four-car suburban EMU stabled in the station; the present-day aspect (below) shows just the two through lines which have been retained, and part of the housing estate which has been built on the ground that was once the station sidings. *R. C. Riley/KB*

CRYSTAL PALACE (HIGH LEVEL) (1): In every way rather more splendid than the LB&SCR station at what was to become Crystal Palace (Low Level), E.M. Barry's Italianate-style LC&DR train shed of iron, glass and yellow and red brick had turreted towers at each corner, passenger accommodation entirely under cover, and a vaulted and tiled chamber as means of access to the Palace from the station; now preserved, the chamber is opened to interested parties for a few days each year by the Crystal Palace Foundation (bottom). In its heyday, the station boasted 33 trains a day in each direction, but as time progressed, its popularity waned, and when the Palace itself was destroyed by fire in 1936, the pleasure traffic dropped almost to nothing and High Level station became run down and delapidated, as shown by the scene taken on 23 October 1954, a month after closure. By March 1990, only the retaining wall remains, with a housing estate having sprung up on the old station site, and the myriad of lines leading to the edifice now a deserted field.
Hulton Deutsch/KB/BM

CRYSTAL PALACE (HIGH LEVEL) (2): Although it was painful to some to witness the closure of an electrified line, traffic levels by 1954 were so poor that there remained no other alternative. The last electric train ran on 18 September of that year, with the 'Palace Centenarian' steam special giving passengers the last opportunity to travel to High Level on the following day. (Above) With the empty stock of this train, two 'C' Class 0–6–0s, Nos 31719 and 31576, emerge from Paxton tunnel and approach the signal box and station. The present-day view of the site, photographed on 1 December 1989, is recognisable only by the retaining wall of the cutting on the right, which has been maintained and refurbished, and the tunnel mouth itself. *R. C. Riley/KB*

UPPER SYDENHAM: Situated within a curve of the Sydenham Hill ridge known as Hollow Combe, Upper Sydenham station was opened in 1884 as part of the Crystal Palace (High Level) line and was intended to serve the well-to-do local population who, for the most part, were housed in grandiose villas within extensive grounds. From the outset, however, the station was almost lost between Paxton and Crescent Wood tunnels (later known as Upper Norwood and Upper Sydenham tunnels respectively), and had to be approached by steps and a steep path down a side of the Combe which was prone to landslips. In consequence, passengers were not a common sight and by 1910 few trains remained which were scheduled to stop there. (Above) With less than a year to go before closure, this view taken on 3 October 1953 shows 4SUB unit No 4292 entering the station. (Below) Just over 36 years later, the house is still in existence and the tunnel mouth can clearly be seen through the undergrowth, providing evidence that a railway once existed here. Inside the tunnel, the walls are encrusted with the soot from steam locomotives of long ago, and even the air inside still smells of the steam railway. *R. C. Riley/KB*

LORDSHIP LANE: (Above) Looking nothing at all like Pisarro's famous painting, Lordship Lane station was opened in 1864 as part of the Crystal Palace (High Level) line, but by 6 July 1954 it looked like this. In fact, closure was to follow within a few weeks and nothing of the railway remains today but the position of the old trackbed, which is now a local walk. *R.C. Riley/KB*

SYDENHAM HILL: (Above) With steam to spare, rebuilt Bulleid 'Merchant Navy' Class 'Pacific' No 35015 *Rotterdam Lloyd* passes Sydenham Hill station's down home signal on 5 April 1959, hauling the Dover-bound 'Golden Arrow' express, with two Mk 1 coaches between the Pullmans and the baggage car. (Below) In the same position on 29 July 1989, the 15.23 Victoria–Dover Priory train is formed of Network SouthEast-liveried Class '423/0' 4VEP No 3156 leading Class '411/5' 4CEP No 1618. The road bridge under which the train is passing remains in situ, but the semaphore signals and background tree have gone and the siding has been taken up. *R. C. Riley/KB*

NORWOOD JUNCTION SHED: (Above) Apart from one BR Standard Class '4MT' 2-6-4T and an Ashford-constructed 0-6-0 shunter, the whole of the Norwood Junction shed (75C) occupancy on 4 April 1962 is made up of Class 'W' 2-6-4Ts and a number of Maunsell 'Moguls'. Serving the extensive Norwood yards, the shed was used to house and maintain freight locomotives and was opened in 1935. Closure came about in 1965 when diesel and electric locomotives replaced steam. Selhurst Depot, south of Norwood station and on the opposite side of the line, remains today, however, having been constructed on the site of the 1926 LB&SCR overhead line depot. (Below) Seen on 30 November 1989, the site of 75C has been taken over by a British Rail Cable & Civil Engineering Supply Depot, the perimeter fence of which, top left, follows the line of the former hump. *R. C. Riley/BM*

ELTHAM WELL HALL (1): Prior to the mid-1960s, most suburban stations had their own small goods yard where a long-winded procedure of 'pick-up' freights operated, with the train locomotive undertaking often complicated shunts in order to get into the yard and couple up to whichever wagons were due for transit. (Above right) On 6 July 1951, 'C' Class 0-6-0 No 31725, of Hither Green shed, awaits a clear road from Eltham Well Hall yard before proceeding to the next station up the line to repeat the procedure. Nothing remained of the yard on 22 July 1989, although its position can still be determined by some of the background houses. *Both BM*

ELTHAM WELL HALL (2): (Left) Rounding the curve into Eltham Well Hall station on 21 January 1983, a Class '508' EMU No 508037 travels towards Slade Green Works for attention. At this time, EMUs of this type were working suburban trains from Waterloo but were later replaced by the Class '455s' and had one trailer removed before being transferred to Tyneside. No sign of Well Hall station now remains; the site is a motorway overbridge for the railway, and a new station, simply called Eltham, has replaced both Well Hall and Eltham Park. *Both BM*

ELTHAM PARK: Opened in 1908 as Shooters Hill station, Eltham Park closed in March 1985, despite energetic objections by the local commuters, and was replaced by a new Eltham station which incorporated both Eltham Park and Eltham Well Hall. It is eminently satisfactory for the Well Hall population to have a nice, new station, but they do not have to walk a quarter of a mile to the new edifice along a fenced corridor open to the elements and, in the opinion of many, not very safe, particularly after the hours of darkness! (Above) This view shows Eltham Park station's original covered walkways from street to platform level with their decorative valance boards and supports, while Class '415/1' 4EPB No 5147 heads the 12.24 Dartford–Charing Cross via Bexleyheath train on 21 January 1983. (Below) On 23 July 1989, only the grass-grown platforms remain as the 11.10 Dartford–Charing Cross service travels past, formed of Class '415/4' 4EPB No 5467 leading '415/1' unit No 5168. *Both BM*

BECKENHAM HILL: The Catford Loop line from Nunhead to Shortlands was opened in 1892 and became an accepted relief route for the LC&DR main line via Penge East, served today by local trains from Blackfriars to Sevenoaks and also by 'Thameslink' services. With Catford being the 1910 limit of the London built-up area, and the Ravensbourne valley above Catford remaining unbuilt over, the station constructed on the line at Beckenham Hill was reputed to be the quietest of suburban stations until the edge of the London County Council's Bellingham estate reached it. (Above) A Ramsgate train heads through Beckenham Hill on 9 May 1959 in the charge of Maunsell 'Schools' Class 4-4-0 No 30938 *St Olave's.* (Below) The same view today is a little obscured as Class '319' No 319046 draws to a halt, forming the 12.11 Cricklewood–Sevenoaks train. The attractive old wooden footbridge on the station has been replaced by a metal one and a block of flats has appeared on the horizon above the houses, but everything else appears to have remained substantially the same. *R. C. Riley/BM*

SHORTLANDS JUNCTION: Rounding the curve between Shortlands and Beckenham Junction at Shortlands Junction, 'Schools' Class 4-4-0 No 30916 *Whitgift* (above) heads for London Victoria with a train from Ramsgate on 27 July 1957, photographed by Dick Riley from the bottom of his garden. Today's aspect is somewhat different, having been widened for four tracks and featuring a signal box, built for the widening, which has since been closed following introduction of the London Bridge power box. Semaphore signalling has disappeared but the sub-station still operates, although it is becoming more and more obscured by foliage in the same manner as the grass bank which made an identical position for a present-day photograph impossible. On 30 November 1989, a Class '411/4' 'Kent Coast' EMU, No 1505, in NSE livery, forms the 12.15 Ramsgate–Victoria service. *R. C. Riley/KB*

BICKLEY JUNCTION (1): (Above) Topping the 1 in 95/100 climb from Shortlands Junction to Bickley Junction on 27 July 1955, Maunsell Class 'N15' 'King Arthur' 4-6-0 No 30769 *Sir Balan* makes little of a ten-coach Ramsgate train travelling on what is today the London-bound track at this point. (Below) With a newly constructed 'wide-shouldered' bridge for Channel Tunnel traffic in evidence, Class '465/0' 'Networker' EMU No 465026 passes on 9 August 1995 forming the 10.35 Blackfriars-Sevenoaks train. In addition to the new bridge, the signal gantry and telegraph poles have disappeared and houses have been built, but the general appearance of this particular stretch of the ex-South Eastern & Chatham Railway remains easily recognisable. *Both BM*

BICKLEY JUNCTION (2): On 8 November 1952 the 10.00 'Continental Express' from Victoria to Dover Marine, carrying passengers for Brussels, comes off the main line at Bickley Junction headed by 'Merchant Navy' 'Pacific' No 35028 *Clan Line*, and commences to round the loop to Petts Wood Junction, which connects the Victoria main line with that from Charing Cross. In connection with the first phase of the Kent Coast electrification of the late 1950s and early 1960s, the loop was redesigned and realigned in order to raise the line speed, and the junction at this point was moved back nearly 100 yards. Then the advent of Channel Tunnel traffic brought about reinstatement of the original loop alignment, which is currently used by 'Networker' EMUs operating the local services, the new 'chord' carrying the 'Eurostar' trains. On 17 November 1995 Class '465/1' unit No 465157 comes off the main line at the same position, forming the 12.38 train from Victoria to Orpington. *Both BM*

ST MARY CRAY–CHISLEHURST LOOP: With a haul of Kent coal, Maunsell Class 'N' 'Mogul' No 31404 traverses the loop which connects the up main line to Victoria with that from Charing Cross. Since the above photograph was taken, on 21 February 1953, the rather mushy grassland bounding the line has become completely overgrown and a very large bramble thicket today prevents coverage from exactly the same spot. (Below) On 12 November 1989, a Class '930' 'Sandite and De-icing' train, No 008, is seen at the same position, formed from a redundant Class '405/2' 4SUB unit. *Both BM*

CHISLEHURST (1): (Above) With Chislehurst station in the background obscured by the exhaust of 'Schools' Class 4-4-0 No 30905 *Tonbridge,* this view of a down Hastings express, taken on 21 April 1951, is identical to the one of Class '202' 6L Hastings units Nos 1011 and 1031 (below) forming the 17.23 Charing Cross–Hastings train over 24 years later, on 19 May 1975. There is no exhaust now to obscure the station buildings, but the branch of a tree which has grown up has achieved the same result! *Both BM*

CHISLEHURST (2): (Above) Looking in the opposite direction from the scenes on the previous page, another 'Schools' locomotive, No 30923 *Bradfield*, steams towards Charing Cross on 8 November 1952, hauling narrow-bodied Hastings line stock, and clatters across the bridge that spans the Victoria main line to the Kent Coast. At the same location on 19 May 1975 (below), all-blue-liveried Class '415' 4EPB No 5367 forms the 18.18 Orpington-Cannon Street train and passes Chiselhurst Junction signal box, which was constructed in connection with the Kent Coast electrification. In common with other modern boxes in the area, this one no longer operates, having been superseded by an all-enveloping power box at London Bridge. The building was finally demolished in 1995. *Both BM*

FARNINGHAM ROAD: (Above) On 11 April 1953, a Victoria–Ramsgate express approaches Farningham Road station headed by 'King Arthur' Class 4-6-0 No 30768 *Sir Balin*, not to be confused with *Sir Balan* depicted on page 79. As expected, the view from the same position on 13 September 1989 (below) reveals that the signal box has gone, together with the pointwork for the sidings and the shunter's cabin; where the station sidings were is now a factory. Class '423/1' 4VEPs, led by No 3471, form the 17.55 semi-fast service from Victoria to Ramsgate. *R. C. Riley/KB*

SIDCUP: (Above) On 24 June 1956, Maunsell Class 'U1' 'Mogul' No 31880 passes through Sidcup station with a freight from Ashford (Kent) bound for Bricklayers Arms. (Below) On 24 November 1988, nothing remains of either signal box or sidings as a pair of Class '33' diesels, Nos 33211 and 33022, are seen at the same spot hauling an up train of 'Murphy Aggregates' from the Isle of Grain. The placement of the discs on the steam engine signified that its train was travelling via the Dartford Loop line; headcode 40 is the present-day equivalent code. *Both BM*

SELSDON: The line from Woodside to Selsdon was opened as a branch of the 'Mid-Kent' line from Lewisham to Beckenham in 1885 and first closed in 1917. Reopened upon electrification in September 1935, it was finally closed again on 12 May 1981 after British Rail had published the statutory notices on the basis that some £500,000 was needed for track renewals, that only some 150 people a day used the line (only 36 at Selsdon itself), and that an alternative station was in close proximity at South Croydon. (Above) On 28 August 1951, the 17.20 train from London Bridge to Tunbridge Wells West arrives at Selsdon hauled by Fairburn Class '4MT' 2-6-4T No 42105 of Brighton shed (75A), the train running on the non-electrified section of track through to Sanderstead. For electric trains to reach Sanderstead direct, a quarter-mile stretch of line through Selsdon was converted, and it is on this section that the 10.07 London Bridge–East Grinstead train is seen, below, on 30 November 1989, formed of Class '423/1' 4VEP No 3463. Although very overgrown, the old station platforms still exist together with the building at the rear of the train, but everything else has been swept away. *R. C. Riley/KB*

GRAVESEND (1): The 'North Kent' line to Gravesend via Woolwich and Dartford was opened by the South Eastern Railway in 1849, the town hitherto having been served by river steamer. Samuel Beazley, architect of London Bridge, designed the principal stations on the line, and the one at Gravesend, together with Erith and Dartford, survives. (Above) On 20 June 1952, the 15.32 service for All Hallows awaits departure time before being propelled by push-pull-equipped Kirtley Class 'R' 0-4-4T No 31658 from Gillingham shed (73D). This service and the old LC&DR locomotive are but a memory, but the young lady in the picture is still sprightly and is seen, below, on 27 July 1989 with her grandchildren, Victoria and Duncan. *Both BM*

GRAVESEND (2): Constructed in a cutting, the line eastwards from Gravesend station to the Medway towns is crossed by nine road bridges, giving the impression of a tunnel. (Above) Running 'light engine' on 20 June 1952, two Maunsell 'Moguls', Nos 31822 and 31856, of Class 'N1' and 'N' respectively, emerge from beneath the arches, returning to their home shed of Hither Green, having worked an unbalanced freight to Hoo. (Below) Apart from the bridge no longer having need of boarding to prevent smoke from drifting across the road and unsighting vehicle drivers, little had changed on 27 June 1987 as empty stock for the 16.26 train to Charing Cross via Sidcup arrives, formed of the expected Class '415/1' 4EPB No 5231. *Both BM*

GRAVESEND WEST: The branch line from Fawkham Junction to Gravesend West Street was opened by the Gravesend Railway on 10 May 1886, became a London, Chatham & Dover Railway branch, and was closed for passenger services by British Rail on 3 August 1953, complete closure coming about on 25 March 1968 after a 15-year life as a goods depot. The initial passenger service was of 14 trains each way on weekdays and eight on Sundays, all through trains to and from London and designed to take advantage of the popular Rosherville Gardens which, previously, were served only by river craft. The popularity of the gardens declined and, following their closure, the line was left to such local traffic as was available. Drastic service cuts were made during the Second World War, leaving just five trains catering for workmen and schoolchildren, and a brief spell as a Continental boat train route for Batavia Line from the pier which extended from the station did not last. As the South Eastern Railway line from Gravesend Central provided a satisfactory service to London, closure was inevitable, and the majority of the station site is now a Great Mills superstore car park (centre), although part of the station and pier structures still stood on 1 August 1989 (bottom). (Top) On 11 April 1953, Wainwright 'C' Class 0-6-0 No 31723 stands beside the dilapidated signal box before backing on to the train in the platform which will form a service to Farningham Road. *R. C. Riley/BM/KB*

ROSHERVILLE: Although the Gravesend West branch was constructed primarily to serve the important leisure traffic from London for Rosherville Gardens, the gardens, in fact, closed for good in 1910 with Rosherville station being reduced to an unstaffed halt in 1928, and finally closing in 1933. (Above) The grass-covered station platform and part of the station buildings were still in place on 11 April 1953 as 'C' Class 0-6-0 No 31713 passed through hauling the 16.45 train from Gravesend West to Farningham Road, but, on 1 August 1989 (below) the site has changed almost beyond recognition. The road bridge over the line, which can just be seen beyond the station canopy in the older view, still exists, but the trackbed beneath has been lowered to allow for road clearance and the old platform is now buried under landscaped banking. The large house in the background remains as a reference point. *R. C. Riley/KB*

SOUTHFLEET: On 11 April 1953, a Gravesend West–Farningham Road train makes the scheduled stop at the island platform of Southfleet, the midway station on the Gravesend West branch, with 'C' Class 0-6-0 No 31713 of Bricklayers Arms shed (73B) in charge. Closed along with the rest of the branch in 1968, nature has quickly taken over the site, and apart from the sleepers still present amid the grass, the station and goods yard are now woodland. *R. C. Riley/BM*

The London, Tilbury & Southend

FENCHURCH STREET: Despite the overhead catenary associated with electrification of the London, Tilbury & Southend services from 1961-62, the small four-platform station in the heart of the City of London at Fenchurch Street still seems to retain much of the atmosphere of days gone by. Opened by the London & Blackwall Railway in 1841 to cater for traffic to and from the developing dock and residential areas along the north bank of the River Thames served only by river steamers, the London & Blackwall joined with the Eastern Counties Railway to form the LT&SR in the mid-1850s, and was taken over by the Midland Railway from 1912. (Above) While still a part of the LT&SR in 1910, a train from Tilbury enters the station hauled by 'Tilbury Tank' No 27 *Whitechapel* and passes another of the class backing into the station, No 34 *Tottenham.* The present-day scene, below, seen on 18 August 1989, shows wider platforms and fewer lines and a complete change in the City landscape to include the Docklands Light Railway terminus structure of Tower Gateway on the right. Entering the terminus and forming the 13.50 from Southend Central is Class '302' EMU No 302211. *R. C. Riley collection/BM*

BROMLEY-BY-BOW: Opened as Bromley station by the LT&SR in 1894 and re-christened Bromley-by-Bow in 1967 in order to avoid timetable confusion with the Bromley stations in Kent, the station was used both by the District Railway and the LT&SR from 1902 until London Transport took over total ownership from BR in 1969. (Above) Photographed from the footbridge, Stanier three-cylinder Class '4MT' 2-6-4T No 42533 enters the station on 4 May 1957 with a Fenchurch Street–Southend Central service. (Below) On 3 February 1989 there is no longer a footbridge on which to take an exactly duplicate view, but the adjacent buildings to the rear remain. LT&SR services now travel non-stop through the overgrown platforms, the only stopping trains being those of London Underground, one of which can be seen in the distance in both views. *BM/KB*

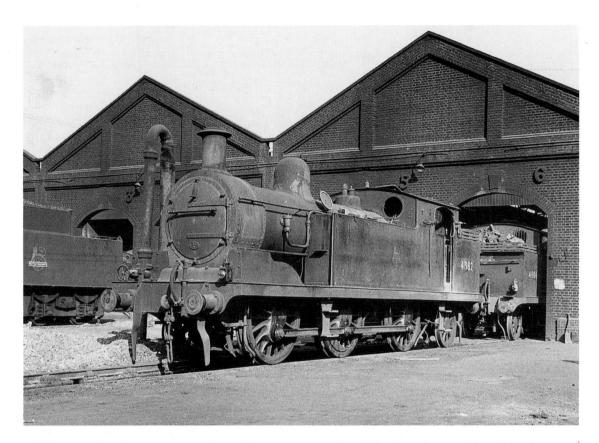

PLAISTOW SHED: The ex-LT&SR shed at Plaistow carried London Midland shed code 13A until coming under Eastern Region control in 1949 and taking up the code 33A. Closed to steam in 1959, the depot became merely a sub-shed to Tilbury (33B) until 1962, when it closed completely. Standing on the exact site of the shed today is the East London Rugby Football Club headquarters. (Above) On 16 April 1955, ex-LT&SR '3F' 0-6-2T No 41982 stands outside the shed building with No 41986 of the same class on the inside. Originally known as the '69' Class, both locomotives once sported names, *Wakering* and *Canvey Island* respectively. *Both BM*

BOW WORKS: The North London Railway works at Bow were fully established as early as 1863 and once covered an area of 31 acres on both sides of the NLR main line. Overhaul of steam locomotives carried on well into the 1950s and work on carriages and wagons did not cease until the mid-1960s. The site was demolished by 1970 and today is solely residential, as indicated by the scene below, taken from almost exactly the same position as the one above, which shows the interior of the works on 16 April 1955 with three types of engine under repair: in the foreground is ex-LT&SR Class '3P' 'Atlantic' tank No 41941 with Class '3F' 'Jinty' 0-6-0T No 47487 behind and Stanier Class '4MT' 2-6-4T No 42514 in the adjoining bay. *BM/KB*

BARKING: (Above) On 23 March 1957, to the east of Barking station, Fairburn Class '4MT' 2-6-4T No 42520 hauls a LT&S lines train from the Essex coast towards Fenchurch Street while, on 1 December 1989, the 12.42 Fenchurch Street–Shoeburyness train is formed of Network SouthEast Sector-liveried Class '302' EMU No 302217. The houses to the right of the line remain, but an additional track has been installed and the yards on the left are now used exclusively by London Underground District Line trains. To the left of the present-day scene, an underpass for London-bound trains has been constructed by London Transport, displacing a number of lines. Semaphore signalling has, of course, disappeared, and ugly overhead electrification masts abound. *R. C. Riley/BM*

PURFLEET: Purfleet was one of the original stations opened by the LT&SR from Forest Gate to Tilbury in 1854. (Above) An afternoon Fenchurch Street–Shoeburyness train draws to a halt on 23 March 1957 powered by BR Standard '4MT' 2-6-4T No 80105. In contrast (below), on 7 July 1989, a similar service is formed of a Class '308' EMU No 308151. The station footbridge and up-side platform buildings have gone and undisturbed foliage appears to be taking over the area above the down side, where the track itself could do with a visit from the 'weed-killer' train. *R. C. Riley/KB*

GRAYS: (Above) On 23 March 1957, Ivatt Class '4MT' 'Mogul' No 43120 of Cricklewood shed approaches Grays station with a boat train for Tilbury Riverside. Chalk outcrops along the north bank of the River Thames brought about the establishment of a lime and cement industry in the area and emissions of the type seen top left were commonplace. Health and Safety regulations resulted in closure, however, and the same chimneys seen behind the undergrowth in the present-day photograph (below) are destined for demolition. The houses alongside the line appear to have lost a number of their chimneys but, otherwise, little change is apparent today apart from the siding with the buffer-stop becoming a through line. Forming the 10.42 service from Upminster to Tilbury Riverside on 1 December 1989 is Class '302' EMU No 302272. *R. C. Riley/KB*

Essex byways

OCKENDON: The 6 miles 53 chains connecting single line from Upminster to West Thurrock Junction, Grays, was opened by the LT&SR on 1 July 1892 and has just one intermediate station and passing loop at Ockendon. (Above) Approaching that station on 19 July 1952, a typical Grays–Upminster train of the immediate post-war period is hauled by Whitelegg ex-LT&SR '2F' 0-6-2T No 41986, whose name, *Canvey Island,* was short-lived, being removed circa 1912. (Below) A present-day branch train at the same location on 5 August 1989, the 16.08 Upminster–Grays, is formed of Class '302' EMU No 302223. *BM/KB*

EPPING: The Central Line of London Underground reached Epping in 1949, the extension having been abandoned pre-war, but the Epping-Ongar services remained steam-hauled until 1957. (Above) On 8 August 1953 the 11.32 Epping-Ongar train awaits departure behind ex-Great Eastern Railway Class 'F5' 2-4-2T No 67213 of Stratford shed, which has just finished watering up. (Below) On 24 July 1995 London Underground 1992 stock forms a Central Line train for West Ruislip; the water column is long gone, the platform has been resurfaced and the chimneys on the station house have received attention. Services to Ongar were withdrawn in 1995, but generally the flavour of a Great Eastern Railway country station is still evident here. *BM/KB*

EPPING SHED: Epping was one of ten sub-sheds attached to Stratford depot (30A). Seen on 8 August 1953, this building dates from 1949, the original 1893 two-road edifice having, almost literally, fallen down of its own volition. Following the arrival of London Underground, remaining steam workings, freights, excursions and the Ongar push-pull trains were operated from here until electrification was completed to Ongar in November 1957 when the little shed finally closed. (Above) On shed on 8 August can be seen two ex-Great Eastern 'J15' Class 0-6-0s, Nos 65444 and 65449, and a push-pull-fitted ex-Great Northern 'C12' Class 'Atlantic' tank No 67363; the two tender engines were used for light freight workings to Temple Mills and Ongar, and the tank engine was tried out on the push-pull trains to and from Ongar but was not particularly successful and stayed only for the one summer. (Below) Nothing now remains of the depot site, the area having become the station car park. *BM/KB*

EMERSON PARK: The 3¹/₂-mile branch line which connects Romford with Upminster was opened in 1893 as part of an LT&SR route to Tilbury via Upminster and Grays; the one intermediate station at Emerson Park opened in 1909 as a result of housing development in the area. (Above) On 19 July 1952, the 14.43 train from Romford to Upminster calls at the station hauled by an ex-Midland Railway '1P' Class 0-4-4T No 58054. With platform extended, changes to the small station building apparent, and a raised bridge constructed to allow for overhead electrification, the 13.53 service from Romford to Upminster (below) runs into the station on 23 February 1989 formed by Class '315' EMU No 315801. *BM/KB*

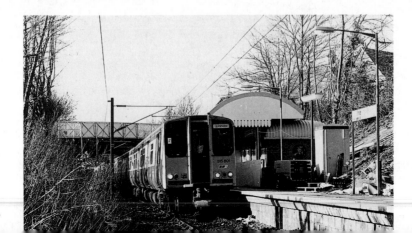

ROMFORD: The present-day Romford station dates from an LNER reconstruction of 1931. The original building was the terminus of the Eastern Counties Railway for one year from 1839 until the line was extended through to Brentwood in 1840. The LT&SR branch to Upminster was opened on 7 June 1893 and was conceived mainly to keep the then rival Great Eastern Railway away from Tilbury, even having its own station entrance until 1934. (Above) Ex-Midland Railway Johnson '1P' Class 0-4-4T No 58038 propels the 12.14 train from Upminster into Romford station on 5 September 1953. Class '315' EMUs now operate the service and the photograph below of No 315801, taken from the same spot, was not made any easier by the forces of nature endeavouring to reclaim the line for their own. *BM/KB*

LIVERPOOL STREET (1): Considered by many to have been the most picturesque of the London termini, Liverpool Street station dates from 1874 and was designed by the Great Eastern engineer Edward Wilson. There have been a number of alterations to the station in the course of its long and busy history, but nothing like the massive redevelopment commenced in 1983 and eventually completed in December 1991. The old pattern of separate areas with long and short platforms has been replaced by a magnificent new station, which retains a greatly renovated Western Train Shed and the Great Eastern Hotel, and now has uniform platforms and eight tracks out as far as Bethnal Green. (Above) The sulphurous interior of the Liverpool Street station steam-hauled suburban lines is graphically illustrated by ex-GER Holden 'N7/2' Class 0-6-2T No 69681 simmering under a May sun in 1951, waiting to depart with a service for Hertford East. The view (below) taken from the same position on 27 November 1988 shows part of the massive concrete raft erected above the suburban platforms, from which Class '305' EMU No 305415 waits to emerge, forming the 16.12 for Bishops Stortford. *BM/KB*

LIVERPOOL STREET (2): (Above) On 10 September 1977, a view towards Liverpool Street station through the washing plant reveals Class '37' No 37025 and Class '47/0' No 47005 at the stabling point awaiting their next duty. (Below) A view from the same position today would actually be beneath the vast concrete raft shown spanning the rails and awaiting a skyward development. On 27 November 1988, Class '47/4' No 47455 backs into the station to take out a parcels train which is completely out of sight in the depths of the interior. *Both BM*

LIVERPOOL STREET (3): The above view of the Liverpool Street main-line departure platforms on 11 April 1951 shows Class 'B1' 4-6-0 No 61250 awaiting departure time with a Norwich express, and No 61264 of the same class backing out of the station following release from the buffer-stops. (Below) From the same position, on 18 August 1989, the camera is beneath the new concrete raft where, in the distance, a four-car Class '312' EMU No 312791 makes up the 13.42 train to Harwich Town. *Both BM*

BETHNAL GREEN BANK: After leaving Liverpool Street station, the lines take almost a 45 degree turn to the right before commencing the climb of Bethnal Green bank at 1 in 70. (Above) The 17.54 Liverpool Street–Norwich express approaches Bethnal Green station on 1 June 1958 hauled by BR Standard 'Britannia' 'Pacific' No 70007 *Coeur de Lion* and passes a Class 'N7' 0-6-2T labouring up the incline with a stopping train for Enfield Town. (Below) From the same position on 5 August 1989, Class '309/3' EMU No 309613 heads for Clacton, passing over white newly-laid ballast. A few of the background buildings remain on the right, but the left-hand side of the scene is completely changed, with tall city offices now on the horizon and the old sidings taken up. *BM/KB*

BETHNAL GREEN (1): On 6 August 1958 (above), Gresley Class 'K3/2' 'Mogul' No 61810 runs tender-first past Bethnal Green station hauling empty stock from Stratford to Liverpool Street, just one of 47 such workings which were required daily at the time. (Below) Running under the same catenary support gantry in February 1989, Class '312/0' EMUs, with No 312709 leading, form a Harwich Town–Liverpool Street service. During the intervening years, the gantry has sprouted a few additional fitments, the down main line platform has been put out of use with an iron railing built down the middle, the station buildings have disappeared and 'modern' flats now overlook the railway instead of the old Victorian homes which were once prevalent. *BM/KB*

BETHNAL GREEN (2): Hauling a Gresley 'Quint-Art' set forming the 18.27 stopping train from Liverpool Street to Enfield Town on 4 June 1958, Gresley Class 'N7/5' 0-6-2T No 69669 makes the mandatory stop at Bethnal Green (above). (Below) On 10 January 1989, the 13.15 local service from Liverpool Street to Cheshunt is formed of a Class '315' EMU No 315813 which draws to a halt at the same platform. Whilst the same platforms are in use, the local lines are now electrified and the original station buildings have been replaced by small brick-built waiting rooms, giving a rather austere look to the present-day scene. *Both BM*

HACKNEY DOWNS (1): (Above) Rounding the curve at Hackney Downs North Junction on 11 January 1958, Class 'N7/3' 0-6-2T No 69686, running bunker-first and hauling a local train for Liverpool Street from Hertford East, shuts off steam for the Hackney Downs station stop. (Below) Apart from the untidy clutter of today's overhead electrification, the junction has not changed as Class '321' EMU No 321313 approaches the station at the same spot on 5 August 1989, forming the 11.22 Cambridge–Liverpool Street semi-fast train. Some of the buildings in the background remain but the majority have either been knocked down or are now obscured by the quite surprising amount of foliage which has grown up here over the years. *R. C. Riley/BM*

HACKNEY DOWNS (2): The North London Line between Hackney Central and Dalston Kingsland stations passes beneath the line from Liverpool Street just south of Hackney Downs station. (Above) On 1 November 1958, a local service for Enfield Town crosses the bridge headed by Class 'N7/5' 0-6-2T No 69663. On 5 August 1989 (below), one is faced with a view which would never have been the subject of a photograph but for the need to depict exactly the same scene! With the accoutrements of overhead electrification providing a particularly ugly foreground, the 12.05 Liverpool Street–Enfield Town train is formed of Class '315' EMU No 315816. *R. C. Riley/BM*

STAMFORD HILL: (Above) With the almost mandatory 'Quint-Art' coaching stock in tow, an equally mandatory Class 'N7' 0-6-2T enters Stamford Hill station on 4 June 1958 with the 14.42 local service from Enfield Town to Liverpool Street. The locomotive, No 69659, passes the neat little lineside hut and telegraph poles which are no longer in evidence but, apart from the increase in foliage, little else has changed in the present-day view (below), which shows Class '315' No 315829 forming the 14.50 Enfield Town–Liverpool Street service on 29 April 1989; one of the background houses has received a white wall-coating and, generally speaking, the occupants of that road now have a more restricted view of the railway. *BM/KB*

STRATFORD WORKS: Apart from Bow Works, which provided a substantial facility for the relatively small North London Railway, Stratford was the one remaining main works not to have moved out of London, and was at the hub of the GER network. Most of the original works building on both sides of the Cambridge lines were demolished in the 1970s, but the Engine Repair Shop, dating from 1915, remains, although no longer in use, having closed in 1991. (Above) On 7 July 1956 Thompson 'L1' Class 2-6-4T No 67740 from Neasden shed (34E) is under repair, together with Class 'B1' 4-6-0 No 61052. (Below) On 30 November 1988, three years prior to closure, an almost identical view of proceedings inside Stratford Major Depot, as it was then known, was possible. Another repair road has been opened up, the premises enjoy girder-mounted heater banks for the staff, and even a rotary brush was then available to keep the floor safe from oil and grease with the operator wearing ear protectors. Where Class '08' shunter No 08958 and Class '47/4' No 47452 are situated was at first used as a locomotive store, but the whole building is currently empty, and awaits an uncertain future. *Both BM*

STRATFORD SHED (1): Working conditions inside Stratford Diesel Depot are considerably more comfortable than they used to be inside Jubilee steam shed, which once stood on the same site; the aroma of diesel fuel now takes the place of the sulphurous, eye-watering atmosphere which prevailed at 30A. (Above) Giving some indication of what the air was like to breath, Class 'K3/2' 'Mogul' No 61830, Class 'B2' 4-6-0 No 61616 *Fallodon*, Class 'K3/2' No 61862 and Class 'N7/1' 0-6-2T No 69624 all await their next turn of duty on 22 August 1954. (Below) In the same position on 1 March 1989 are Railfreight Sector-liveried Class '31/1' No 31238, Class '37/0' No 37057 and a pair of Class '08' shunters, Nos 08627 and 08533. *Both BM*

STRATFORD SHED (2): On 10 March 1956, Hill Class 'Y4' 0-4-0 dock tank No 68128 commences to haul trucks away from one of the Stratford coal lines, ground loading having been accomplished by crane. (Below) Although still rail-connected, the site today sees more road vehicles than trains, being part of an industrial complex which has been constructed there. *Both BM*

STRATFORD SHED (3): Stratford Works once had its own small two-road shed for housing the Departmental engines used for works shunting. Constructed just before the First World War, it could house four tank locomotives under cover and was not finally demolished until 1981, although it had been out of use for a considerable period prior to that. (Above) On 29 January 1955, Class 'J66' 0-6-0Ts Nos 31 and 32 await use alongside Class 'F5' 2-4-2T No 67218. By 30 November 1988, only part of the concrete base remains amid the encroaching grass and shrubs. *Both BM*

STRATFORD SHED (4): (Above) Taking the line west of the Stratford shed complex on 10 April 1954, Class 'B1' 4-6-0 No 61006 *Blackbuck* hauls a transfer freight from Temple Mills across the wastes and approaches High Meads Junction. (Below) Only the terraced houses in the right distance provide a focal point for the present-day scene taken on 19 January 1989 with Class '31/1' No 31263 running round its train of petroleum tanks. Stratford Freightliner Terminal is situated behind the concrete road flyover on the left and the line is now electrified as part of the freight-only spur from the North London Line through Copper Mill Junction and the remains of Temple Mills yard to Channelsea North Junction. *BM/KB*

TEMPLE MILLS YARDS: (Above) Taking up the slack of a loose-fitted mixed freight for Colchester, Worsdell ex-GER Class 'J15' 0-6-0 No 65370 begins to move slowly away from Temple Mills yards on 10 April 1954. Adjoining a water-mill that was once owned by the Knights Templars, Temple Mills marshalling yards evolved from sidings in 1877, through a series of extensions to 1930, when the 'Cambridge' hump yard was constructed. The layout was modernised between 1954 and 1959 and during this period there were 74 daily freight arrivals at the south end with 51 departures from the north. A small area of the complex remained open after the main activities of the yards ceased, and even today a large expanse of the area remains unused as can be seen in the view from the same position in 1988. However, closure of Leyton Civil Engineering Depot to enable construction of the contentious M11/A12 road link has resulted in Temple Mills being rejuvenated and put to use as a Permanent Way Depot and On-Track Crane Depot at a cost of £4½ million. Opened at the end of 1994, the new depot's prime user is BR's East Anglia Infrastructure Services, with train movements to and from Temple Mills depot provided by Mainline Freight Ltd. *BM/KB*

FOREST GATE: Opened by the Eastern Counties Railway in 1840, Forest Gate station was expanded by the GER and later by the LNER in 1946, when electrification brought about the re-opening of the station after its closure for most of the war years. (Above) On 8 August 1953, a down express working is unusually in the hands of Gresley Class 'K1' 'Mogul' No 62070, which appears in good fettle as it approaches Forest Gate with steam to spare. (Below) On 27 March 1989, little has changed as the 14.35 Liverpool Street–Kings Lynn express is seen from the same viewpoint powered by Class '86/2' electric locomotive No 86260 *Driver Wallace Oakes GC. BM/KB*

BRENTWOOD BANK (1): For 2 miles preceding Brentwood, the ruling gradient is 1 in 103 with the steepest section at 1 in 99, quite a climb for a heavy steam train. (Above) Climbing the bank on 5 September 1953, Class 'B17/6' 'Footballer' 4-6-0 No 61655 *Middlesbrough* slugs towards Ingrave summit with a train from Liverpool Street to Clacton. (Below) Seeming to pay the ruling gradient scant concern, Class '309/3' 'Essex Express' EMU No 309616, in Network SouthEast Sector livery, heads another Clacton-bound service on 22 April 1989. With seven stops on the way, this one takes 1½ hours for the journey, an improvement of 30 minutes on the days of steam haulage. Comparing the two photographs, only the overgrown embankment edges, the loss of a siding and the lighter paintwork on the catenary supports appear to indicate the march of time, although the Class '309s' no longer operate on Great Eastern metals, these trains having been taken over by Class '321s'. *BM/KB*

BRENTWOOD BANK (2): (Above) Scampering down Brentwood bank from Ingrave summit to Brentwood and Warley station on 5 September 1953, Class 'B1' 4-6-0 No 61006 *Blackbuck* displays express headcode lamps as it heads a Liverpool Street-bound train consisting of suburban compartment stock. (Below) In exactly the same position on 22 April 1989, Class '321' 'Dusty bin' EMU No 321330 forms the 13.02 service from Southend Victoria to Liverpool Street. With the line to Shenfield having been electrified in 1946, catenary supports were prevalent even in steam days, resulting in little change between these two scenes taken 36 years apart; the retaining walls to the cutting are somewhat blackened by the passage of time although, conversely, the bridge carrying the A128 road has been brightened up with fresh paintwork. *BM/KB*

SHENFIELD (1): Although electrification between Liverpool Street and Norwich was not completed until 1987, the first stage to Shenfield took place in 1946, culminating east of Shenfield station, where the route to Southend divides. (Above) At the point where the wires ended, 'Britannia' 'Pacific' No 70003 *John Bunyan* leaves the Southend lines behind, hauling an express from Liverpool Street to Norwich and Yarmouth on 10 May 1952. (Below) With overhead electrification now continuing and the Southend line equally energised, Class '86/2' No 86229 *Sir John Betjeman,* in InterCity Sector livery, passes the same spot on 22 April 1989 hauling colour-matched coaching stock forming the 11.30 service from Liverpool Street to Norwich. *BM/KB*

SHENFIELD (2): The Great Eastern Railway did not manage to reach Southend until over 30 years after the LT&SR had arrived there, a branch from the main line at Shenfield eventually opening in 1889. (Above) Rounding the curve to Mountnessing Junction, between Shenfield and Billericay, on 10 May 1952, Class 'B17/6' 'Sandringham' 4-6-0 No 61646 *Gilwell Park* sports the distinctive Southend line discs of the day with a service from Liverpool Street. (Below) An equivalent Southend Victoria train at the same position on 22 April 1989 is formed of Class '312/0' and '312/1' EMUs Nos 312718 and 312798. The banking has been cut back during electrification work and the trees have had 37 years in which to grow, but the pleasantly rural scene has not significantly changed despite the catenary masts. *BM/K*B

SHENFIELD (3): Coming off the main line at Shenfield Junction, the down 'Southend Loop' is a bi-directional single line as far as Mountnessing Junction where it joins the two tracks illustrated on page 123. (Above) On 10 May 1952, a Liverpool Street–Southend Victoria express uses the loop and is shown here travelling beneath the Norwich main line hauled by Holden Class 'B12/3' 4-6-0 No 61546. (Below) At the same location on 22 April 1989, the 10.24 Liverpool Street–Southend Victoria train is formed of Class '321/3' EMU No 321327; the 'P' after the unit number indicates that the seats tip up at this end to allow for Parcels Sector use. The iron railings on the bridge have been replaced by brickwork and the once pleasant embankment is now completely covered with brambles. *Both BM*

The 'Widened Lines'

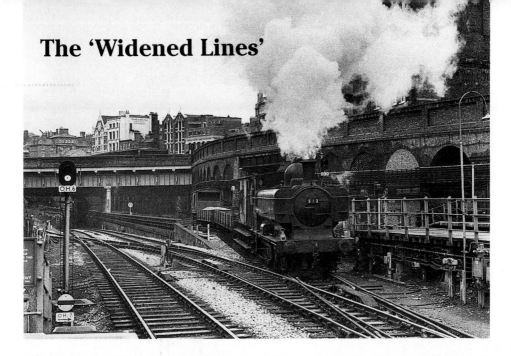

FARRINGDON: During the 1850s, London streets became jammed with slow-moving horse traffic and a journey by bus was liable to be as slow as one in the rush hour today. Plans were therefore laid for the railways of the day to connect together their northern termini and bring the City within easier reach of both the northern and western suburbs. The first section of what was to become the Metropolitan Railway was opened in 1853 and ran from Paddington to Farringdon Road (now Farringdon). Two single-track tunnels connected with the GNR at King's Cross, a junction was opened with the LC&DR at West Street, near Farringdon, and a triangular layout was completed in 1871 by a spur from Aldersgate Street (now Barbican). These connections formed the only north-south route across the City, and heavy traffic resulted in the Metropolitan constructing a second pair of tracks, always known as the 'Widened Lines', between King's Cross and Moorgate. By 1880 there were some 200 trains a day over the Widened Lines into Moorgate, and another 100 over the LC&DR link through Holborn Snow Hill. Competition from London's new tubes and trams, however, saw the virtual demise of the services, only peak-time trains from the Midland line to Moorgate remaining, with even the dieselised GNR trains having ceased in 1976 and the King's Cross tracks taken up. The decline of the route was totally reversed, however, with the reopening of the Snow Hill link and the advent of today's 'Thameslink' trains. (Above) London Transport's GWR-style 0-6-0 pannier tank No L90 crosses from the Metropolitan to the Widened Lines on 26 April 1970 with a Sunday engineer's train. (Below) With the connection no longer in existence, a longer platform and a new brick wall, a Class '319' 'Thameslink' EMU No 319013 arrives at Farringdon on 7 January 1990 forming the 11.22 train from Bedford to Sevenoaks. *Brian Beer/KB*

KING'S CROSS (YORK ROAD): The Widened Lines connection with the GNR at King's Cross consisted of the East Branch on the up side and the Hotel Curve on the down side. (Above) Prior to plunging into the East Branch tunnel on 12 June 1975, Class '31/1' No 31192 stops at the York Road platform to allow passengers to alight. (Below) The same viewpoint on 9 December 1989 shows a few of the many changes in track and platform layouts that have taken place outside the station. Although the Widened Lines have been taken up here, their facility still exists, and a link to King's Cross was planned to be restored as part of the erstwhile King's Cross/St Pancras redevelopment, which appears to have been abandoned now that the new Channel Tunnel terminus is to be at St Pancras and not King's Cross itself. *Both BM*

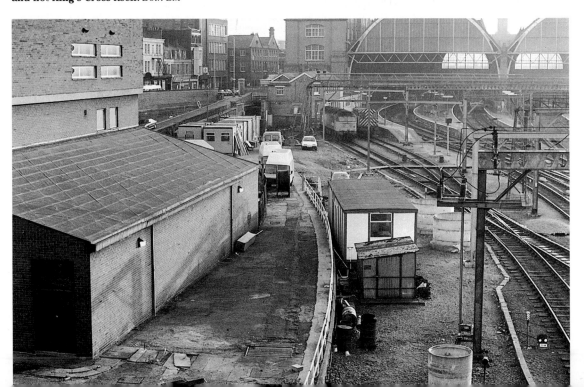